I0800016

Sound Actions

Conceptualizing Musical Instruments

Alexander Refsum Jensenius

The MIT Press
Cambridge, Massachusetts
London, England

The MIT Press would like to thank the anonymous peer reviewers who provided comments on drafts of this book. The generous work of academic experts is essential for establishing the authority and quality of our publications. We acknowledge with gratitude the contributions of these otherwise uncredited readers.

This book was set in Sabon by Westchester Publishing Services. Printed and bound in the United States of America.

Library of Congress Cataloging-in-Publication Data
Names: Jensenius, Alexander Refsum, author.
Title: Sound actions : conceptualizing musical instruments / Alexander Refsum Jensenius.
Description: Cambridge : The MIT Press, 2022. | Includes bibliographical references and index.
Identifiers: LCCN 2021062387 (print) | LCCN 2021062388 (ebook) | ISBN 9780262544634 (paperback) | ISBN 9780262372206 (pdf) | ISBN 9780262372213 (epub)
Subjects: LCSH: Music—Performance—Psychological aspects. | Music—Psychological aspects. | Musical instruments. | Musical perception.
Classification: LCC ML3838 .J38 2022 (print) | LCC ML3838 (ebook) | DDC 781.1/1—dc23/eng/20220415
LC record available at https://lccn.loc.gov/2021062387
LC ebook record available at https://lccn.loc.gov/2021062388

10 9 8 7 6 5 4 3 2 1

Dedicated to my father,
Jørgen H. Jensenius (1946–2017).

Contents

Acknowledgments

Many people have been of great importance to this project. Initial sparks of inspiration came from Arnt Inge Vistnes, who introduced me to the Fourier transform and laid the ground for my interest in studying music from a scientific perspective. Jøran Rudi and Bjarne Kvinnsland introduced me to the world of electro-acoustic music at Notam and helped with my first music technology projects. Tellef Kvifte sparked my interest in organology, while Jon-Roar Bjørkvold and Even Ruud helped shape an understanding of the importance of embodied and cultural perspectives on musicking.

Over the years, I have enjoyed working closely with Rolf Inge Godøy at the University of Oslo. He instilled in me a curiosity about musical sound and, particularly, the thinking of Pierre Schaeffer on sound objects. This led us onto a joint adventure of exploring music-related motion, on which we have later collaborated in many different contexts. As an exchange student at the University of California, Berkeley, I was fortunate to work with computer music pioneer David Wessel. He introduced me to timbre research and artificial neural networks, but even more importantly, he showed me the joys of performing with live electronics. As a visiting Ph.D. fellow at McGill University, I was introduced to motion capture by Marcelo M. Wanderley. I also learned a lot about new interfaces for musical expression and the importance of rigorous testing and evaluation.

I am grateful for all the funding I have received throughout the years. The Norwegian Research Council has supported several of the projects within which the ideas for this book project were developed: Musical Gestures, Sensing Music-related Actions, and Human Bodily Micromotion in Music Perception and Interaction. I should also mention the importance of the EU Cost Actions Gesture Controlled Audio Systems (ConGAS) and Sonic Interaction Design (SID), which helped build up a network of like-minded scholars around Europe. More recently, I have benefited from collaborations in the Nordic Sound and Music Computing network project funded by the Nordic Research Council.

In addition to my wonderful international colleagues, I am also fortunate to work in a music research "powerhouse" at the University of Oslo. I always learn a lot from discussions with colleagues and students at the Department of Musicology. Over the last few years, I have been fortunate to be part of building up the RITMO Centre for Interdisciplinary Studies in Rhythm, Time and Motion with colleagues from the Departments of Musicology, Psychology, and Informatics. Together, we aim to expand our understanding of rhythm as a fundamental property of human life. Thanks to all of you; none named, none forgotten.

Finally, thanks to my mother, Grete, for commenting on the manuscript and giving visual guidance on the many figures in the book, to Valentina and Esperanza for showing me the joys of learning to play new instruments and bearing with me when I bring home all sorts of weird instruments for them to try out, and to Paula for continued support and encouragement.

Prelude

"Hey, I want the tablet now." For my postmillennial daughters, playing music on a digital tablet is as commonplace as playing on an acoustic instrument. They are the first of a digital generation, having had access to all sorts of electronic devices their entire life. I have given them plenty of possibilities to test various instruments: all kinds of "normal," acoustic instruments, but also different devices that produce sound "electro-acoustically." Such instruments are often called "electronic" or "digital," but they do, in fact, produce acoustic sound. Therefore, I prefer the term electro-acoustic for all instruments that run on electricity. I also use a hyphen between electro and acoustic to emphasize that these instruments contain both electronic and acoustic parts.

Electro-Acoustic Instruments

The term "electro-acoustic" may sound archaic to some. When I last did a raise of hands among a group of students, the response clearly showed that the term is not within their vocabulary. A few thinks about "tape music," in which prerecorded music is played over loudspeaker orchestras. In performances of "electroacoustic music" (usually written without a hyphen), the loudspeakers are often thought of as the "instrument." This may seem odd to some. Still, the loudspeakers do produce the acoustic sound that one hears. While working on this book, I have gradually realized that "electroacoustic instruments" is the most precise term for describing all instruments that generate acoustic sounds based on electronic sound generation. It does not matter whether the instrument is a classic analog synthesizer or a mobile phone app. They both share the property of producing audible sound with some electronic circuitry; hence, they are electro-acoustic instruments.

Electro-acoustic instruments are ubiquitous these days. Many toddlers start their musical explorations with battery-driven, colorful, plastic-made musical toys. According to my definition, these are electro-acoustic instruments. But how does a battery-driven percussion instrument differ from an acoustic drum? After pondering such questions for years, it has become clear that acoustic and electro-acoustic instruments are fundamentally different in design and construction. They also "afford" different types of musical expression. That is, their design and construction invite specific musical actions. To generalize, one could say that acoustic instruments appropriate intimate "sound" making. On the other hand, many electro-acoustic instruments open for interactive music making. There are no rules without exceptions, and, as we shall see in later chapters, there are both examples of acoustic music makers and electro-acoustic sound makers. Still, I believe some fundamental differences exist between instruments that produce sound mechanically and those that require electricity. To progress in our reflections on musical instruments, I think it is necessary to acknowledge these differences.

This book is, in many ways, an attempt to understand my musical journey. I played mainly acoustic instruments in my childhood. Then I went on to study piano at university—first classical, then jazz, then experimental music. I was also interested in technology and gradually began exploring computer music. My entry point into electro-acoustic musicking came from the digital side. It was first after ten years of digital music making that I discovered the joys of analog synthesizers. This route is not uncommon for my generation, growing up when the personal computer appeared on the market. The previous generation started with analog electronics, and the current generation grows up with mobile devices. These cultural differences are, of course, significant in shaping our musical selves. At the same time, these three generations—analog, digital, and mobile—share the experience of making music electro-acoustically. We are the first generations for which electronically produced music has become the norm.

Embodied Music Technology

Parallel to my interest in designing, building, and performing with various types of instruments, I have spent the last two decades researching within the emerging field of "embodied music cognition." Here the central idea is that the body is an active participant in both music performance and perception. This may seem obvious, but acknowledging the importance of the body is, in fact, a relatively new and radical idea within musicology (Clarke 2005; Leman 2008). My main argument in this book is that musical experiences should be understood as interactions between human bodies and musical instruments. Most acoustic instruments are played through a close, physical connection between the body of the performer and the body of the instrument, what I will call an "action–sound coupling." In comparison, electro-acoustic instruments are based on "action–sound mappings" that can be more or less "disembodied."

There is a growing understanding of the importance of relationships between action and sound in music. Emerson and Egermann (2018, 106) argue that a better "gesture–sound causality" leads to greater understanding and appreciation:

> A higher degree of perceptible causality, as created by the mapping and the type of controller, provides spectators with more information and more reference points for evaluating the performance. Being able to perceive, understand, and then create a mental model of the sound generation process not only generates greater interest, it also appears to provide a basis for assessing the amount of skill displayed in the performance.

I aim to understand more about such causality by analyzing what I will call the "action–sound separation" afforded by musical instruments. As I will argue, the action–sound separation of acoustic instruments is, by design, lower than for electro-acoustic instruments. This does not mean that acoustic instruments are necessarily "better" than electro-acoustic. A well-tuned, high-quality acoustic grand piano will undoubtedly outperform a piano app on a mobile phone in terms of sound quality. However, the app may have other qualities. For example, it can teach you music theory and inspire you to play. Therefore, it does not make sense to compare a grand piano to a piano app. While they share some attributes, they are examples of two instruments with entirely different designs and musical affordances.

As a music technologist, I want to develop new and better technologies, both acoustic and electro-acoustic. As a performer, I am primarily concerned about the feeling of the instruments I play. As a musicologist, I am interested in understanding how instruments influence the people performing and perceiving them. Over the years, I have seen the need for a set of theoretical "tools" to talk systematically about instruments: established knowledge, empirical evidence, well-defined terminology, and conceptual models that can be applied in musical discourse. This book is my attempt at creating such a set of theoretical tools, building on theories from relevant fields.

In some disciplines, theoretical modeling comes first, followed by experimentation and, finally, practical applications. The field of music technology can be described as a meeting point between practice-based disciplines: art, design, and engineering (Wang 2018). Few music technologists engage in theoretical development, and not many music theorists have a background or interest in music technology. Thus, most music technology books fall into two categories: (1) technology-focused how-to guides or (2) cultural-historical narratives of how particular technologies have shaped music. Some of the influential books in the field are either more geared toward describing techniques (Roads 1996) or focused on electronic music (Chadabe 1997; Holmes 2002; Collins and d'Escriván 2007). There are some influential music cognition books that touch upon technological issues (Leman 2008, 2016), and some technological books with a cognitive touch (Cook 2002). Many books focus on particular parts of the field, such as musical interface design (Miranda and Wanderley 2006), programming techniques (Boulanger and Lazzarini 2011), or cultural perspectives (Butler 2014). Numerous anthologies have been published in recent years (Bovermann et al. 2017; Ruthmann and Mantie 2017; Holland et al. 2019), but while they present good overviews of ongoing research, they often fail to provide a coherent theory. This book follows some other recent monographs that aim at developing new music theories, such as those by De Souza (2017), Wang (2018), and Magnusson (2019). Their stories are more closely related to philosophy, interaction design, and composition, respectively. My book can be seen as a meeting point between interactive music technology and embodied music cognition. I am particularly building on recent research in embodied interaction (Dourish 2001; O'Neill 2008; Streeck et al. 2011) and embodied music interaction (Lesaffre et al. 2017). If I should call my approach something, it would be "embodied music technology."

The Conclusion First

This book has a straightforward argument: relationships between actions and sounds are at the core of music making. Consequently, such relationships should form the basis for thinking about music. A little more than a century ago, Edgard Vàrese famously defined music as "organized sound" (Risset 2015). This definition paved the way for a new way of thinking about music as sonic art. In fact, over the last century, music has become so sound focused that the body is sometimes forgotten. My alternative embodied definition would therefore be that music is "organized sound-producing actions." This definition builds on the action–sound theory that I will present in the book. My argument can be summarized as follows:

1. Music is both an active and embodied process. There is no such thing as a passive "listener;" everyone takes part in the act of "musicking" with all senses. New technologies allow for different types of musicking than before, making it necessary to reconsider traditional musical roles, such as those of composers and performers.
2. Acoustic sounds are produced by the interaction of physical objects, what I call an "action–sound coupling." In most acoustic instruments, there is a small "action–sound separation," which means that there is a close interaction between the performer and the instrument. More technologically advanced acoustic instruments often have a larger action–sound separation than "simpler" instruments.
3. Electro-acoustic instruments are based on designing interactions between actions and sounds: "action–sound mappings." Such mappings often have a larger action–sound separation than acoustic instruments. Electro-acoustic instruments tend to embed more musical knowledge and structure, becoming more of a "music maker" than a "sound maker."
4. A higher "spatiotemporal resolution" in electro-acoustic instruments may help decrease the perceived action–sound separation. On the other hand, instruments composed of many separate modules allow for a larger "spatiotemporal distance" between an instrument and its performer. This challenges the idea of performing "here" and "now."
5. New "hybrid instruments" bridge the gap between acoustic and electro-acoustic instruments. Acoustic instruments can be expanded with music-making electronics. It is also common to add acoustical and mechanical components to electro-acoustic instruments. Such experimentation paves the way for the future of musicking.

The argument will be built on various existing theories from relevant fields. I will also reflect on my own musical explorations with different instruments. As such, the book will move between objective descriptions and a subjective narrative.

Techno-Cognition

I will use the term "techno-cognition" to describe the cognitive processes involved when a user interacts with technology. This is something else than studying only human cognitive processes or a device's technical operations. Kockelman (2013) describes techno-cognition as the "cognition of technology," such as how logic gates and algorithms sense the world. My use of the term covers this perspective, but I also include humans in the interaction chain. After all, performing on an instrument is based on the continuous *inter*action between the human body and the instrument. As such, it does not make sense to evaluate an instrument as a static object. It needs to be studied in use and with a human as part of the interaction chain.

A more speculative part of my argument is applying a techno-cognitive perspective for those who perceive sound-producing actions. This builds on our remarkable cognitive abilities when it comes to identifying relationships between actions and sound (Yost et al. 2008). Human sound-source perception is based on the ability to "hear" the sonic results of actions that we only see or to "see" the actions of sounds we only hear. We are even able to both "see" and "hear" actions and sounds entirely from memory. This you can check by performing a simple thought experiment: Think about a glass falling toward the floor. By imagining this in your mind, you can "see" the glass crashing into pieces when it hits the floor and "hear" the ringing sound of glass pieces breaking. A more musical example would be to think of a violinist performing sustained bowing. Even though you may never have played the violin yourself, you can probably come up with both audible and visual "images" in your mind of the performance. Also, if you have played yourself, you may "feel" the tension of holding the violin in your left hand and the bow in the right.

Despite increasing knowledge about human cognition, basic principles of action–sound couplings are often violated in electro-acoustic instruments. Sometimes they are violated deliberately, resulting in exciting musical experiences. My research into "inverse instruments" is an example of this. As will be discussed in chapter 12, these instruments only produce sound when you intentionally do *not* play them. The experience is bewildering, yet many people find it exciting. In other cases, mappings may only be confusing. For example, think about how generic MIDI controllers are used with any sound generator. Keyboards are based on on/off messaging, which works well for piano-like sounds. Such binary control does not work equally well for sustained instrument sounds, like that of a violin. Hence, the performer will be unable to control the sound continuously. This is confusing for the perceiver, who will not understand the relationship between the sound-producing action and the resultant sound. The result is an action–sound mapping that violates basic physical and cognitive principles. My argument is that a techno-cognitive approach may help in designing better electro-acoustic instruments.

Interdisciplinarity

One of the great things about writing a book, as opposed to conference papers or journal articles, is the opportunity to think broader and deeper. While working on this project, I have tried to make connections between fields that are often surprisingly separate, such as music sociology, music psychology, music technology, and music aesthetics. One may think that these fields have many things in common and are informed about each other's concepts, theories, and models. In my experience, they follow remarkably separate research trajectories. I have no ambition of creating an overarching theory that works well within all of these fields. However, I will try to combine some concepts from the different fields with some of my own.

My approach is genuinely interdisciplinary. However, what does this entail? Stember (1991) proposed definitions for the following types of disciplinarities:

- **Intradisciplinary:** Working within a single discipline.
- **Crossdisciplinary:** Viewing one discipline from the perspective of another.
- **Multidisciplinary:** Working together with people from different disciplines, each drawing on their disciplinary knowledge.
- **Interdisciplinary:** Integrating knowledge and methods from different disciplines, using a real synthesis of approaches.
- **Transdisciplinary:** Creating a unity of intellectual frameworks beyond the disciplinary perspectives.

She argues that many researchers that claim to work interdisciplinarily actually work multidisciplinarily. I like to think of the different types of disciplinarities along a continuum with a gradually increasing level of integration, such as sketched in figure 0.1. This book project started on the cross- and multidisciplinary side and has moved toward inter- and transdisciplinary integration. At some point, one could imagine that full integration occurs.

Identifying the target group is challenging when working interdisciplinarily. When writing within a discipline, one may assume familiarity

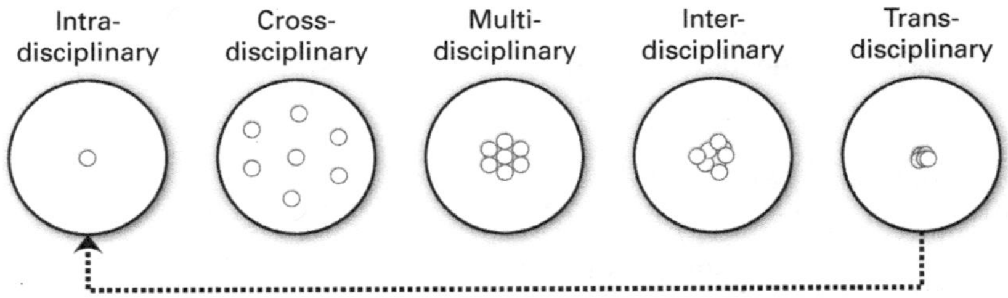

Figure 0.1
A sketch of different levels of disciplinarity, combining the taxonomy of Stember (1991) with the drawing by Zeigler (1990). While drawn as separate categories, the different levels should be seen as following each other on a continuum.

with core concepts and ideas. Throughout the years, I have presented theories from this book to many different audiences. It is remarkable just how different the responses are when I talk in front of groups of, say, musicologists, psychologists, or computer scientists. It is equally challenging to target groups of composers, performers, or designers. I have seen the need to "translate" terms and explain concepts to fit the various audiences. I try to do the same in this book.

Artistic versus Scientific Research

In addition to working between different scientific disciplines, my work is also situated in the arts. I like to think of this as a continuum, as sketched in figure 0.2. Unfortunately, art and science are often institutionally separated, although there are some examples of how art and science are connected and even integrated (Borgdorrf 2012). No doubt, there are different theories, methods, praxes, and outcomes between art and science. However, in my experience, artistic methods can be used to develop new knowledge, and scientific methods can lead to new artworks. Some examples of such intertwined knowledge and art production will be presented in chapters 11 and 12.

The second axis in figure 0.2 is between the humanities (culture) and the natural sciences (nature). Within universities, these traditions are often split into separate faculties. The natural sciences are usually hypothesis driven, while the humanities are based on hermeneutic interpretation and reasoning. However, there are numerous examples of how humanities scholars use empirical and experimental methods while natural scientists reason and infer. There is also an increased understanding of the need to work in cross-faculty teams to solve the big societal challenges. However,

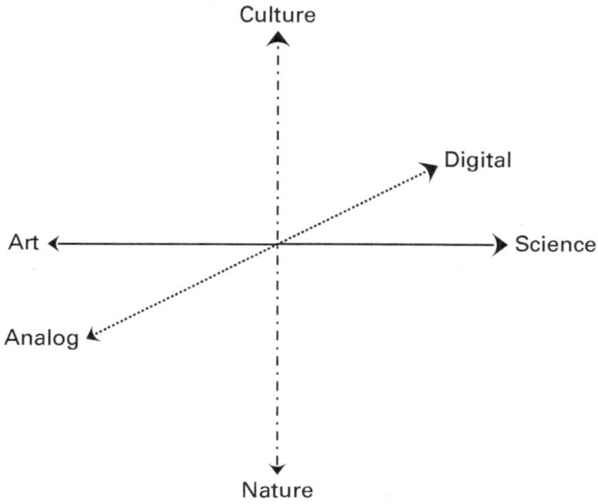

Figure 0.2
This interdisciplinary book project is placed between three (independent) axes: the study of objects along a nature–culture axis, outputs along an art–science axis, and technologies along an analog–digital axis.

in my experience, that is often easier said than done. Working across faculty borders is challenging at best and impossible at worst. That is probably why most "interdisciplinary" projects end up being cross- or multidisciplinary.

In the third axis in figure 0.2, I draw up a separation between analog and digital technologies. There is a tremendous focus on digital technologies these days. Nevertheless, we should not forget all the analog technologies surrounding us, such as door handles, knives, swimming suits, and guitars, to name a few. We should also remember that digital devices have analog components to interface with the analog world. For example, a mobile phone captures and produces sound, light, and vibration. That is why I find it fascinating to investigate the similarities and differences between analog and digital technologies.

Outline of Book

The book is organized into four parts, each presenting one of the project's four main topics:

- **Musicking:** Thinking of music as an active process ("to music"), and how this changes our ideas of instrument making, composition, performance, perception, and analysis. This is also closely connected to questions about musical embodiment and the role of instruments in musical experiences.

- **Embodiment:** Defining core terminology from (bio)mechanics and understanding different functions of music-related body motion: from the sound-producing actions to communicative gestures of performers to various types of entrainment and sound-accompanying motion of perceivers.
- **Interaction:** Investigating the design and construction of both acoustic and electro-acoustic instruments, and the techno-cognitive differences between action–sound couplings and mappings, their action–sound separation, and the importance of spatiotemporal resolution.
- **Affection:** Reflecting on the aesthetic, embodied, and emotional experiences of playing various instruments. Here I will present some of my own prototype instruments and reflect on their design, construction, and performance.

It should be possible to read the individual parts of the book independently, as they each introduce a complete set of conceptual building blocks. To understand their connections, however, it will be necessary to read the book linearly. I also recommend trying out various instruments while reflecting on the presented concepts. After all, musicking is an active and embodied process.

I

Musicking

1

Music as an Active Process

"Can I try?" the old man asked as he looked up at me from below the stage. I had just completed a performance with some of my music balls, a collection of ball-shaped instruments that I will present in more detail in chapter 11. "Yes, of course," I replied and quickly turned on the devices again. "I used to play the piano as a child," the old man said before he quickly added, "but I stopped since I was no good at it." More people approached as he carefully lifted and tested the sonic capabilities of each of the music balls. For about half an hour, I had a group of audience members standing on stage. The music balls are not the most advanced instruments, but they look nice and make unconventional sounds. Up until that point, my main focus had been on building instruments for myself. The experience showed me that many people want to engage more actively with music, but they are afraid to "not be good at it."

Musical Engagement

There is music everywhere—at home, in the streets, in shops—yet most of that music is experienced "passively" through various playback devices. The musical perfectionism made available through modern-day music production tools has raised the bar for what we expect from a musical performance. Even the best artists and musicians cannot deliver the same sound in a live performance as they did after multiple takes and adjustments in the recording studio. This discourages many from engaging in music making. Yet, as DeNora (2000) describes, musical experiences are central to many peoples' everyday lives.

One day, my youngest daughter came home from school and told me that her teacher refused to sing with the children because "she couldn't sing." Instead, the teacher had turned on a recording of a children's song and asked the children to sing along. As my daughter recalled, most children sat on their chairs and only listened to the music. They, too, said that "they couldn't sing." How did we get to a point where children believe they cannot sing?

The development of new music technologies has changed how music is performed and perceived over the last century. Before the twentieth century, people experienced music by playing or singing themselves or being in the close vicinity of someone else performing. Today, many people have access to all the world's music from their mobile phones, yet they may never touch an instrument or sing themselves. In one way, one can say that new technologies have led to a musical "democratization" process (Théberge 1997). Consumer technologies and particularly new mobile devices have given people access to consume music everywhere. It is paradoxical, then, that we simultaneously see such a high level of musical "passivization."

The challenge with many traditional instruments is that they are hard to master. As documented by Bjørkvold (1992), all children engage in sing and play. Many also start to play an instrument. Unfortunately, many also quit after a relatively short time. Each year, around half of the new children at my daughters' school join the marching band, but most of them leave within a year. Many say that it is too difficult, that they do not have time to practice, or that they are "no good at it." One could say that they should persist, but that is easier said than done. Not everyone wants to spend the time and effort to learn an acoustic instrument. Yet, the alternative does not have to be that they should never engage actively with music making again.

While I worry about the decline of singing in schools, I am optimistic about the future of music in general. The experience with the music balls taught me that there is a need for musical instruments that afford "easy" musical engagement. Fortunately, there is a lot of ongoing research in the field. I am involved in a music technological subdiscipline that investigates the development of new instruments, or what is often referred to as NIMEs after the annual International Conference on New Interfaces for Musical Expression (Jensenius and Lyons 2017). This community researches everything from specialized expert devices to "active listening" technologies. Examples of the latter are music players that include remixing features or collaborative music games. There is much potential in developing new technologies with a complexity level somewhere between traditional instruments and music playback devices. That is what I will call "musicking technologies." We will get back to many examples of such technologies later. First, I will introduce the concept of "musicking" and connect it to my techno-cognitive reasoning.

Musicking

The term "musicking" was introduced by Small (1998) in his book *Musicking: The Meanings of Performing and Listening*. His main argument is that engaging with music should be considered an "active" process. In the book, Small (1998, 9) introduces the verb "to music" as

> *To take part, in any capacity, in a musical performance, whether by performing, by listening, by rehearsing or practicing, by providing material for performance (what is called composing), or by dancing.* We might at times even extend its meaning to what the person is doing who takes the tickets at the door or the hefty men who shift the piano and the drums or the roadies who set up the instruments and carry out the sound checks or the cleaners who clean up after everyone else has gone. They, too, are all contributing to the nature of the event that is a musical performance.

Musicking is a broad concept. It covers everyone involved in a musical activity, both directly and indirectly. Small argues that musical meaning is not something within the music "itself," but something that arises from the engagement with music. This message can be seen as an attack at the mainstream musicological discourse of the time, which was focused on canonized works by dead, white, male composers. On a side note here, I should clarify that I will be using "musicology" in its etymological meaning, that is, the study of music. While in some countries musicology has become more or less synonymous with "music history" (Kerman 1985), I am using musicology as the umbrella term for all the academic subdisciplines studying music: history, theory, sociology, philosophy, psychology, technology, and others (Parncutt 2007).

Considering his overt criticism of classical music-focused musicology, it is paradoxical that Small (1998) spends a large part of his book on an analysis—or, perhaps more precisely, a dissection—of a traditional classical concert hall experience. This analysis includes everything from the concert hall's foyer layout to how the musicians and conductor walk on stage. The concert is seen as a ritual in which everyone present plays a part. This is also true for the audience members, what he calls the "spectators," who usually sit in dead silence during the performance, follow strict rules for when to change position in their chairs between movements, and only clap at the end. However, these things change. The silence of a contemporary concert hall audience is a relatively new thing. Mozart and Beethoven were used to more lively audiences during their time (Small 1998, 44).

While useful in getting his message through, the controversial nature of Small's argument and his confrontation with the rituals of Western art music may be reasons that his general ideas—and the term musicking—never really caught on outside some musicological subdisciplines. There are mainly two subdisciplines that have embraced the musicking concept: music education and music therapy (Ansdell 2014). A reason for this may be that these fields share an interest in looking at music as a "process," not as a final product. For example, in music therapy, the musicking concept adequately describes the interaction between therapist and participant. The aim is often to improve communication and interaction skills and the general psychological well-being of the participants. Thus the musical interaction is more important than the musical content. Then the concept of musicking fits nicely, as it focuses on the process of making music together. However, what is the role of the music therapist and client? Are they performers or perceivers? I think the question is ill posed; a music therapy session is an example of collaborative musicking.

Rereading Small's book twenty years after I first encountered it, his argumentation feels less controversial. After all, there have been several other attempts at changing the direction of mainstream musicology over the years, including feminist musicology (McClary 1991), empirical musicology (Clarke and Cook 2004), embodied musicology (Leman 2008), and popular musicology (Scott 2009). Even though the term "musicking" may not have been embraced in mainstream musicological discourse, there has undoubtedly been a widened reflection on musical engagement at large.

Although the concept of musicking has been with me for years, I have never followed up on it in previous publications. One reason for not using the term was to avoid stepping into what seemed to be a somewhat sensitive area between those that use the term as a cornerstone of their academic thinking and those repelled by its usage. Since it is a lesser-known term, I often felt it would be too complicated to introduce and explain it rather than just write the same meaning differently. The main reason, however, was that I never really needed such a word before. As I have been developing some of the conceptual building blocks for this book project, Small's ideas have come back to me and helped focus my thoughts. I have found it liberating to incorporate the concept of musicking into my thinking. It feels genuinely inclusive. Rather than separating the traditional notion of performer and perceiver—and the related processes of performing and perceiving—it covers both of these processes (and more) with only one word.

My "tweak" of the term includes adding a techno-cognitive perspective to the act of musicking: studying the role of instruments in music performance and perception. This was not a usage that Small had in mind, but it logically follows his idea of thinking about music as an active process. This way of using the term also fits well with new ideas in the field of embodied music cognition, in which the body is seen as an active participant in the formation of meaning in music.

The Musicking Quadrant

At the core of the reasoning in the rest of this book is what I will refer to as the "musicking quadrant," a schematic layout of different types of musical engagement. You may have a sneak peek at the musicking quadrant in figure 1.8 right away, but it will probably make more sense after I have gone through its different elements in the subsequent sections.

The musicking quadrant should be understood from a subjective point of view: an individual's creation of and experience with music. It simplifies a complex reality, but its multilayered construction still allows for a systematic approach to evaluating the different roles and processes found in musicking contexts. In this chapter, I will primarily focus on the quadrant from the perspective of traditional (acoustic) musicking. Later, it will be used for investigating the complexities of electro-acoustic musicking.

The musicking quadrant contains three layers: (1) the different roles, (2) the actor's type of musical engagement, and (3) the temporal unfolding of the processes. In the following, I will go through these layers one by one before putting them together in the complete model.

Different Roles

One way of thinking about musicking is to focus on the people involved and their different roles with the music. A traditional musicological approach to evaluating the different roles in (Western art) music is proposed by Goehr (1992, 102–103):

> Within classical music practice we compose works, produce performance of works, appreciate, analyze, and evaluate works. To do this successfully we need a particular kind of general understanding. Every time we talk about individual musical works we apply this general understanding to the specific case.

Here the focus is on the musical "work," a traditional entity in the study of music, an amalgam between the composer's musical ideas and an (idealized) performance rendering. What is missing from Goehr's list of musical

Table 1.1
Some roles of musicking and their processes and products.

Role	Process	Product
Instrument maker	Building instrument	Instrument
Composer	Creating musical idea	Piece/tune/work
Performer	Performing piece/tune/work	Performance
Perceiver	Experiencing performance	Experience
Analyst	Analyzing musicking	Analysis

roles—and that is rarely mentioned by other musicologists either—is the role of instrument makers. This comes as no surprise, as instrument maker is a profession that generally receives little attention. Many people think of instruments as mass-produced objects. Then there is a long distance from the designer's original idea to the product that reaches the end-user. Things are different for people that can afford handmade instruments. For them, instrument building is often considered a craft. A good violin maker is respected but rarely credited. Few people would think about the art of instrument building as similar to the art of composition or performance.

It is easy to forget that today's traditional instruments have gone through many experimentation cycles. It is fascinating to wander through instrument museums and look at the many strange examples of violins, brass instruments, and pianos invented over the centuries. However, the art of instrument making is rarely acknowledged in the history of music. There are some notable exceptions, including some famous seventeenth-century Italian violin makers and some legendary twentieth-century synthesizer manufacturers. In most other cases, the instrument maker is a "forgotten" actor in the musical ecosystem. Consequently, the instruments they make are usually not considered a musical artwork on its own either, even though instruments are a necessity for music making. Since instruments are at the center of my attention, my model includes the instrument maker as a core actor. Therefore, I suggest five musicking types, each of which contains actors, processes, and products, as summarized in table 1.1.

In a traditional, Western context, the roles and processes typically follow a linear, "feed-forward" order, as sketched in figure 1.1. The model starts with the instrument maker designing and building the instrument. The instrument maker may work alone, such as an expert guitar maker.

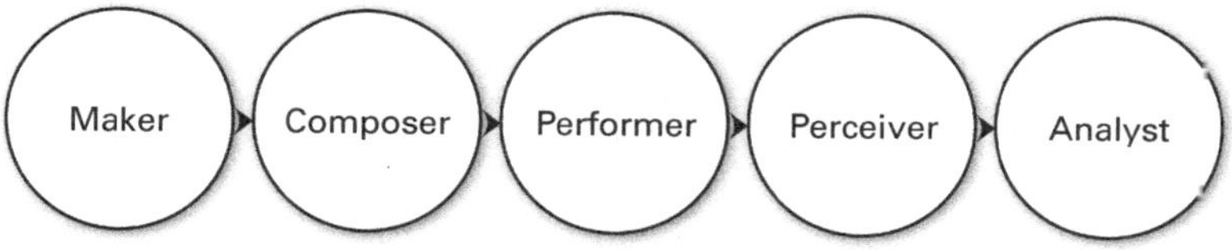

Figure 1.1
A linear, "feed-forward" process typically found in traditional musicking.

Nowadays, the instrument maker's role has been industrialized, such as in a piano factory. Then it is possible to identify subcategories within the instrument-making chain: initial design phase, prototyping, user testing, manufacturing, marketing, and so on. The number of people involved in making an instrument does not matter from the model's point of view. Some musicians make instruments themselves or have a working relationship with a craft-based instrument maker. In most cases, however, people buy a mass-produced instrument without thinking much about how and by whom it was built.

The second role is that of the composer, the person that creates the piece to be played. This role may or may not be closely related to the third role, the performer, since composers may also play their own music. In some musical genres, particularly in those involving improvisation, the composition and performance processes may be wholly intertwined, what is sometimes called "comprovisation" (Dudas 2010). The same is true in many popular music styles, in which the music may be composed and performed by an artist or collectively by a band. From my simplified model's point of view, it still makes sense to include composition and performance as two separate processes. Separate study programs and professional organizations attest to composition and performance being seen as distinct roles in the music world.

The fourth role is often called "listener," but I will refer to it as perceiver. This is to stress that musicking is an embodied experience in which all modalities come into play. Listening—with the ears—is essential, but the sound of music is only part of a musical experience. As will be discussed more later, the other senses are also important, including seeing, smelling, and touching.

The fifth role is that of the analyst, which includes the process of thinking and reflecting about musical experiences. This role should not be constrained to academic investigations of music. Any type of retrospective musical activity could fall into this category, including remembering a concert. Thinking about music—systematically or unsystematically—involves "musical imagery," what Godøy and Jorgensen (2001) refer to as the ability to "hear" sound for your inner ear. This is something we all do, and which in itself may give rise to strong musical experiences.

Even though I have deliberately tried to separate the roles here, they certainly overlap. Some people may take on all the roles, albeit not at the same time. Others may take on two or three roles. As such, the roles are not mutually exclusive. Splitting them is an analytic approach to understand more about their inner workings. This will come in handy when we start investigating various types of instruments later.

Music Production

To complicate things a little, I could add a sixth role: the music producer. Burgess (2013, 5) defines music production as the "technological extension of composition and orchestration." Thus on one side, we could consider the role of a producer as similar to that of a composer. That may be true for the role of many producers today. However, the first music producers were primarily doing live recordings of performances. Such recordings were made on the fly, and the recordists role could be considered close to that of a performer. They worked in real time and only had one take. Today, we find a large variety of music producers, ranging from those running large-scale studios with many people involved to do-it-yourself singer-songwriter making complete albums in their bedroom (Jones 2020). There are also the producer-DJs going on stage, mostly working as performers (Kjus and Danielsen 2016).

To generalize, I will, in the following, use music producer as an umbrella term to cover all aspects of what is often thought of as the "art of record production" (Frith and Zagorski-Thomas 2012). This includes the engineering-oriented tasks of recording, mixing, and mastering. It also includes handling people and making artistic decisions at all stages throughout the process. Today's music producers do not only passively record other's music; they are central in creating musical structures and shaping its "sound" (Brøvig-Hanssen and Danielsen 2016). In figure 1.2, I have, therefore, placed them somewhere between composers and performers.

The role of music producers often overlaps with the other roles. Sometimes they have a distinct role alongside composers and performers. Other times, they may have double (or triple) roles as composers and performers. Still, it makes sense to separate the music producer's role from the

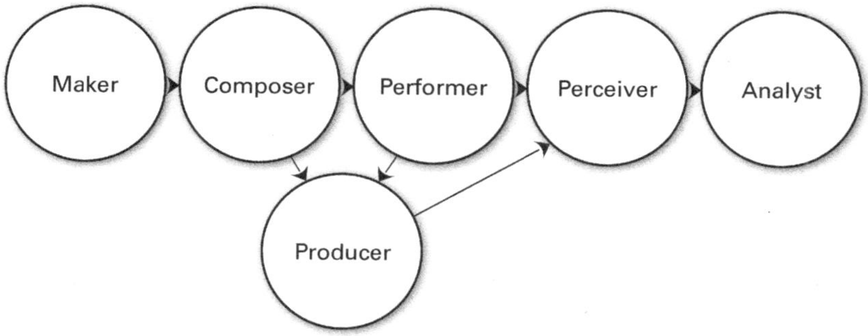

Figure 1.2
The music producer can be placed between composers and performers in the chain.

composer and the performer. After all, they also have education programs and separate professional organizations and follow different career paths.

Creation versus Experience

One way of separating the different musicking roles is to see whether the activity in question is related to creative practice. I deliberately avoid talking about "active" and "passive" musicking here since one of the core elements of the musicking (and embodiment) paradigm is that everyone is actively engaged in one way or another. That is not to say that there are no differences between creating and experiencing music within a traditional musical context. I suggest thinking of them along a continuum, with creation on one side and experience on the other. Figure 1.3 shows how it is possible to place the different actors in a grid with the experience–creation continuum on one side and time on the other. Here the temporal axis refers to when activities happen with respect to the "now" of a performance.

In this system, the instrument maker, composer, producer, and performer are placed on the creation side since they all create something (instrument, piece, performance, recording). A traditional perceiver and analyst will, in this system, be on the experience side. It is crucial, however, to think of one's placement along the continuum as dynamic. For example, a musician may perform during only one part of a concert while perceiving in others. Audience members may sit or stand still during parts of a performance but may also take on a more active role, such as singing or clapping. Understanding such dynamic role changes is essential to explain the complexities of real-world musicking.

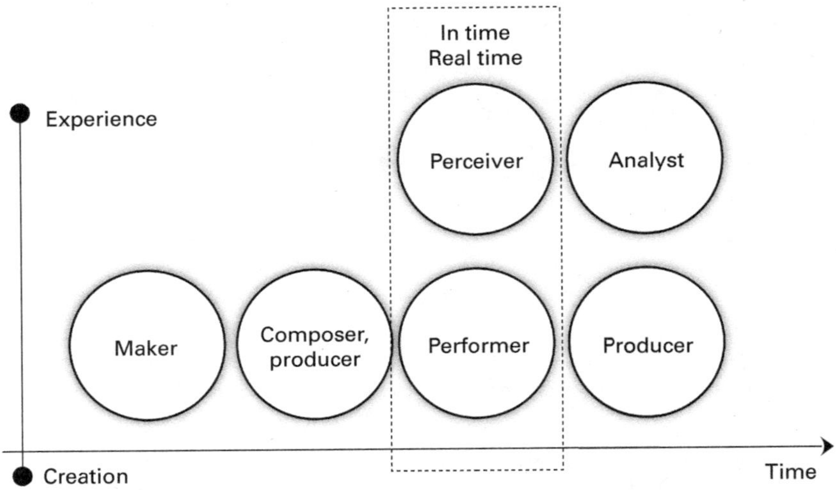

Figure 1.3
The different musicking roles may lean more toward either experience or creation. The temporal axis refers to the order of events with respect to the "now" of a performance.

Temporal Awareness

Another way of looking at the different musicking roles is concerning time. Much has been written about how we experience music in time. See, for example, the works by Kramer (1988), Barry (1990), London (2004), Savage (2017), and Kozak (2019), to mention a few. My approach to musical time is shaped by a techno-cognitive perspective, combining phenomenological and technological models of time.

Dainton (2018) proposes to group different models of temporal awareness into three categories:

Cinematic model: Our awareness is similar to that of taking "snapshots" of the continuous unfolding of time. An example is how the rapid projection of still images is perceived as a "moving image."

Retentional model: Our awareness contains episodes of consciousness. These episodes contain a combination of the direct experience and representations ("retentions") of the past.

Extensional model: Episodes of an experience have a temporal extension and can be seen as extended "chunks" of time.

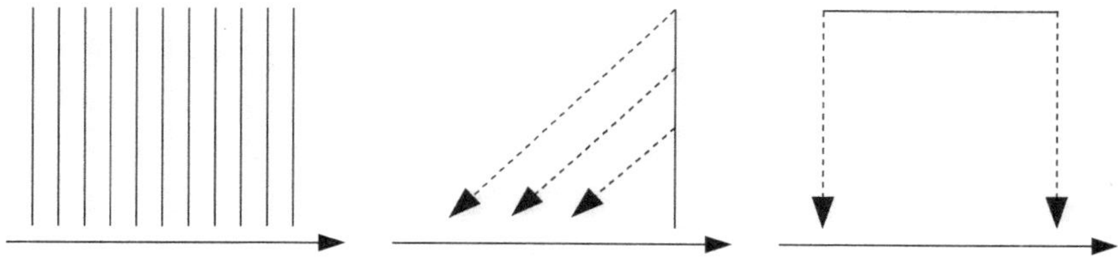

Figure 1.4
A sketch of the three main models of time consciousness: cinematic, retentional, and extensional, adapted from Dainton (2018).

As sketched in figure 1.4, the cinematic model can be seen as similar to how computers handle time. Digital computers sample the continuous data flow at regular time intervals. Each sample represents a moment in time without reference to its surroundings. Temporal fluidity in the playback of such independent samples is ensured through a high-enough sampling rate. This way, our senses are "tricked" to believe that we watch or listen to a continuous phenomenon, even though the actual representation is a series of individual images or sound samples. Standard sampling rates—25 fps for video and 44.1 kHz for audio—are just about fast enough to "fool" our visual and auditory systems.

The retentional and extensional models can be seen as ways of preserving a memory of the past. In the retentional model, one may think of what Husserl (1991) called "primal impressions" of a "now-point" in time. These primal impressions are always connected to what happened ("retentions"). They also anticipate what will happen in the future ("protentions"). In a discussion of the experience of the sound of an approaching coach, Husserl (1991, 290) writes:

> If we focus reflectively on what is presently given in the actually present now with respect to the sound of the postilion's horn, or the rumbling of the coach, and if we reflect on it just as it is given, then we note the *trail of memory* that *extends* the now-point of the sound or of the rumbling. This reflection makes it evident that the *immanent thing* could not be given in its unity at all if the perceptual consciousness did not also encompass, along with the point of actually present sensation, the continuity of fading phases that pertain to the sensations belonging to earlier nows.

In Husserl's model, the retentions are not the same as other types of memory; they are merely the gradually "fading away" of past "nows." Using a signal processing metaphor, this may be thought of as a decaying delay line. What was once at the forefront of the signal gradually disappears.

Much has been written about the philosophy of time consciousness, but I will here make a jump to a more recent psychological model proposed by Stern (2004). As a psychotherapist, his focus is on creating tools he can use in a clinical setting. His concept of the "present moment" can be seen as a combination of the retentional and extensional models. The aim is not to explain the continuous unfolding of time, but rather to create a tool for talking about "nowness" in a clinical setting. A core feature of the present moment is that it has an extension in time. This temporal extension is typically up to around three to five seconds, which is the approximate limitation of our short-term memory (Snyder 2000). A present moment is short, yet at the same time rich, in content. For Stern, it is also essential that the present moment is a whole—what is often called a "gestalt" in the psychology literature—rather than being decomposed into smaller units. A present moment has both a "now" and a "here." This is something we will get back to in discussions about spatiotemporal distance in chapter 9.

Stern uses the present moment as a core element of his therapeutical practice, getting clients to talk about and analyze such moments from memory. He also reflects on how the present moment is active in musicking (Stern 2004, 29):

> ... a musical phrase contains an immediate past and future. The form of the musical phrase is revealed and captured by the listener as the crest of the immediate present instant passes from the still resonating horizon of the past ... toward the anticipated horizon of the future.... Much of the charm of listening to music lies in the surprises that the composer provides by inventing final paths that surprise us but do not overly violate the implications we sensed.

If we "zoom" closer into the experience of music, Godøy (2008) has suggested that we may not necessarily need to choose between the different awareness models; they may coexist. He proposes to think of the perception of an unfolding temporal stimulus as a combination of continuously and discontinuously updated "chunks," as sketched in figure 1.5. This may work because it is not the beginning and end of the chunks that are of importance, but their "goal points." In music, such goal points can be individual tones or combinations of tones that form a chunk—for example, a musical phrase.

From a techno-cognitive perspective, it is interesting to reflect on how a computer's time handling compares to the models of time awareness. As discussed above, sampling can be seen as similar to the snapshots of the cinematic model. However, recent digital representations are based on techniques similar to those of the retentional and extensional models. For example, video compression methods are based on storing "keyframes"

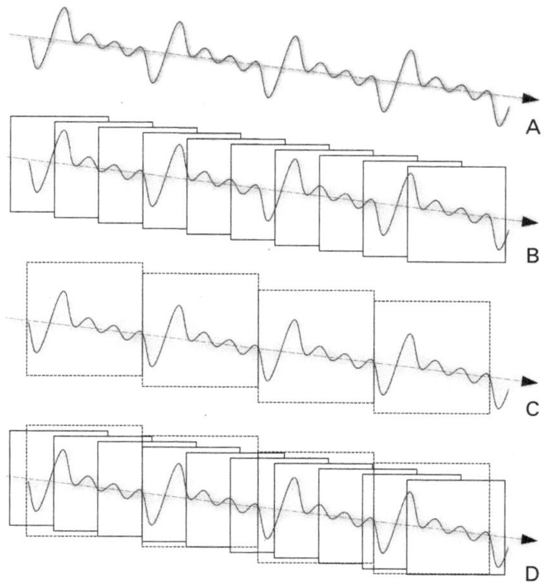

Figure 1.5
Three different types of awareness (B–D) based on a continuous sound signal (A): a floating time window (B), a discontinuous chunk-based updating (C), or a combination of B and C (D), based on Godøy (2008).

at regular intervals. These keyframes contain a complete image, while the following frames only contain information about the pixels that have changed from the keyframe. This is based on the technique of "frame differencing," meaning that subsequent frames are subtracted from each other. The resultant image only contains information about the changes between frames and can be used for motion analysis (Jensenius 2018). Such a keyframe-based compression technique results in much smaller file sizes than if all the images had to be stored entirely. We find similar methods used in compressed audio formats, such as MP3 and AAC, which use psychoacoustical models in the different temporal filters to reduce the size of audio files (Brandenburg 1999). These multimedia formats are good examples of how knowledge about human psychology and cognition can help create better and more efficient technologies.

The Speed of Time

One aspect that we have not touched upon so far is the speed of time. This may be an odd topic from a cognitive point of view, but it is relevant when dealing with computers in which the processing speed is often a significant limitation. Broadly speaking, we can think about two temporal levels in computing systems: "real time" and "non–real time." A real-time system is based on continuously taking in some input, performing the programmed calculations as quickly as possible, and then sending out the result. This is the way most electro-acoustic instruments work. From a cognitive perspective, it may be more relevant to talk about what Wöllner et al. (2018) refer to as "clocktime." However, when balancing between cognitive and technological concepts, I have found it more useful to use the term real time, since it will primarily be used when discussing technological issues.

Since a chain of operations necessarily takes some time, there is always some delay in a real-time, computer-based system. The delay between a system's input to its output can be measured as the system's latency. Sometimes the latency is below a perceptual or cognitive threshold, in which the technological imperfection does not matter from a perceptual point of view. Other times, the latency is evident to the perceiver but still tolerated. After all, we are used to compensating for various types of latencies, and, as argued by Sethares (2007), several of them are also the basis for various perceptual phenomena and musical features (figure 1.6). There are also cases in which latency is a significant problem, which we will discuss in chapter 9.

The perception of latency is an example of the human ability to experience that something is not in time with the current "now." For example, think of a thunderstorm. The lightning can be seen seconds before hearing the thunder. Still, you know that the thunder and lightning come from the same action, even though the vision and sound are experienced at different times. If you only hear the thunder, you will immediately think that you did not see the lightning. Moreover, when you see the lightning, you will wait to hear the thunder. But when does the lightning/thunder actually happen? At the time you saw the lightning? When you heard the thunder? How do they relate to your sense of it happening in the "now"?

Let us move to a more musically relevant thought experiment. Imagine seeing someone pressing a key on a digital synthesizer without hearing any sound. One second later, you hear a sound. If no other sound-producing actions are made, you may perceive that the action and sound are related yet with a perceivable latency. What if the latency was one minute? Or one hour? Unless you were at an experimental concert during which the performer would wait for the sound with the audience, such sonic delays would be far beyond your latency tolerance. Wessel and Wright (2002) suggested ten milliseconds as the optimal target latency when developing electro-acoustic instruments. However, few computer-based instruments

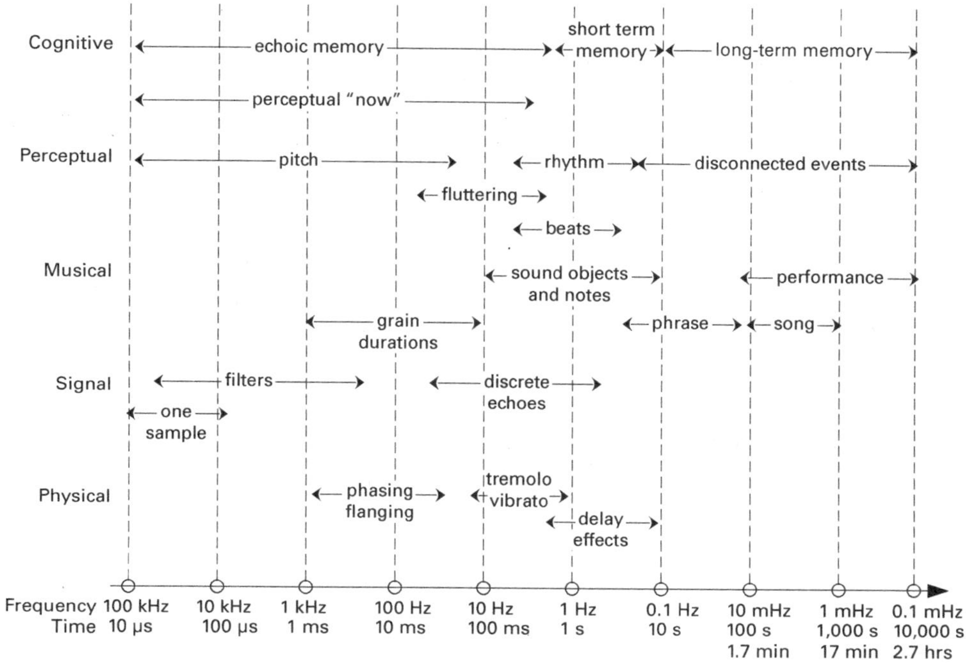

Figure 1.6
A schematic overview of different temporal scales, adapted from Sethares (2007, 7).

can meet such a "10-millisecond" requirement in practice. Yet, many instruments are still perceived as having tolerable latency levels. As illustrated in figure 1.6, whether something is experienced as immediate or not in a musical context highly depends on the musical content.

Even though our latency tolerance may vary depending on content and context, it is quite clear that we have a sense of something being "in time" with the experienced "now." Some events can also be considered "out-of-time," as sketched in figure 1.7. This can be an action that happened in the past. Alternatively, it can also be a prediction or planning of something that will happen in the future. Scheduling is essential when creating computer-based systems, as they allow for preparing various actions that can be triggered based on the incoming signal.

When computers process in time, we typically want them to also work in real time. For example, if you set up a microphone with a reverb effect for a performance, you would like it to work both in time and in real time. However, if you just want to add reverb to a prerecorded song, it does not matter when or how fast it happens. You may probably prefer that the computer does the processing "faster than real time." If the computer can

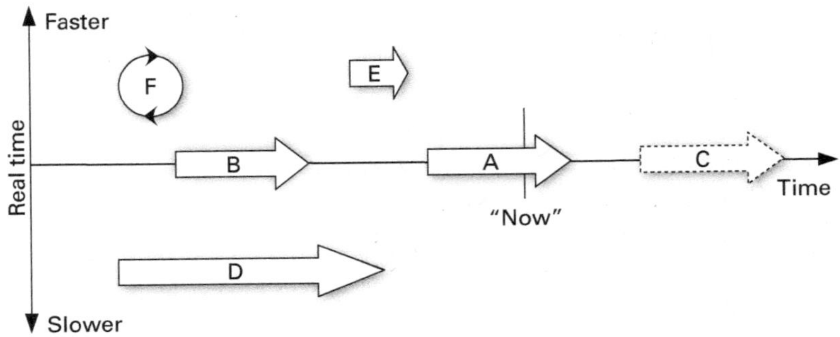

Figure 1.7
Relationships between different temporal levels. An event can happen in the "now" both in time and in real time (A). We may also think of an event that happened in the past (B) or predict a future event (C). It is also possible to create slower (D) and faster (E) events with machines. Both conceptual and technological musical events may also be based on nonlinear time—for example, circular time (F).

process a three-minute sound file in seven seconds, it is unnecessary to wait three minutes to finish its processing. You may also happen to have a computationally heavy reverb model that runs "slower than real time." For a producer working in a studio, it is not crucial whether a process runs in real time. For a performer on stage, however, there is no other option than running in real-time mode. The concert is happening in the "now," hence the tools used need to run both in time and in real time. Since there are so many different use cases and requirements, and time is so critical in music, it is not strange that computer musicians talk a lot about latency issues. This is also one reason why there are many approaches to handling time in music programming languages (Dannenberg 2018).

Kozak (2019, 3) argues that musical time can also be thought of independently from "objective" time:

> Imagine time differently, then—as a sphere, or a cube, or even a hexacosichoron. Imagine it running diagonally, or folding back upon itself, or sideways, or from the inside out. Imagine time crackling, wheezing, rustling, swooshing, buzzing. Imagine time as silent. Now imagine it smelling of freshly cut grass, or a musty hotel lobby. Then again, what if time glistened and shimmered? What if it breathed, slowly, in-out-in-out-in-out? What if it came near you, so close that you could feel its warmth, embrace it, hold it in your hands? What if it did all of that at once?

The idea is that musical time does not exist independently but is based on the interaction between musical sounds and moving bodies engaged in musical activities. This is an exciting line of thought and one that I would like to follow more in the future. However, in the context of this book, I will think of time in a more traditional sense. Still, given the techno-cognitive perspective, I will consider time as both objective and subjective at the same time.

A discussion about time—whether something is happening in time or out of time, in real time or non–real time—only makes sense when we have a reference point: the "now." The "now" is a subjective and relative entity. One might imagine that agreeing on a "now" should be easy for basic musical structures. However, recent rhythm studies have shown that even identifying the "beat" in individual tones is a nontrivial task (Danielsen et al. 2019). The task is much more difficult when one deals with complex music. I remember a concert in which a guitarist used a looper in one of the songs, a device that can record and play short snippets of sound. When the performer came on stage, it was not immediately clear that she was using a looper since it was just one of many boxes in her setup. I was, therefore, puzzled when I could see her playing on the guitar without any sound coming out of the speakers. The guitar sound was routed into the looper, which recorded the first layer of the tune she was about to play. When she stepped on the looper's pedal to stop recording, the sound started playing. Here a temporal disruption of action and sound lead to a sense of an out-of-time performance. This was based on my reference point being the "now" of her performance action. A looper plays with the repetition of the "nows" of the past. Later in the same concert, the guitarist also started playing with the speed of the different sounds she recorded into the looper, making them faster and slower. This could be seen as the in-time use of real-time and non–real-time processing of the sound.

The case of the looping guitar is focused on relatively short time scales, but we may extend the ideas of in time and real time. Let us look at the roles of the musicking quadrant concerning a traditional concert with acoustic instruments. Then it is natural to consider that both the performer and perceiver are involved in in-time and real-time musicking. If we use the "now" of the concert as the reference, then the performer's rehearsal of the concert program earlier the same day would be considered out of time. Similarly, one audience member's imagery of the performance (in the form of a mental "replay") later that day would also be considered out of time. Similarly, we can argue that the instrument maker's role in making the instrument used in the performance is out of time concerning our imagined performance.

A composer's work would generally be considered out of time. The composition process would usually be slower than that of the final musical performance (slower than real time). However, sketching out the form of an extended composition may happen quickly (faster than real time). The composition process may also occur in real time if the composer improvises a piece that is later written down and reperformed in concert. One such example is *The Köln Concert* by Keith Jarrett, first performed in 1975 as extended piano improvisations. After popular demand, the concert was later notated and released as authorized sheet music (Jarrett 1991). What was at first an improvised performance is now played as a "composed" piece of music.

The temporal dimensions of the analyst's role are equally complex. It is common to start an analysis process as a perceiver. Still, music analysis is often carried out both out of time and slower than real time concerning the performance one is studying. This could involve detailed listening and relistening of certain parts of a piece of music to understand its harmonic or textural changes.

The complexity of all possible temporal combinations can be high. My interest is not to complicate things too much but rather to explain the general differences between in-time and out-of-time processes. This is also useful when expanding the model to include music production. For example, a live recording may be considered a separate (in-time) layer between the performer and the perceiver. However, as sketched in figure 1.2, modern-day music production is often a much more circular process between the composer, performer, and producer. In such cases, the temporal complexities of a multilayered model are high, with different layers of in-time and out-of-time processes overlapping. Such complexities are particularly noticeable when it comes to "live" music. As noted by Danielsen and Helseth (2016), many pop music concerts are based on the playback of prerecorded elements. As such, the liveness of the performance is built around the idea of creating a "mediated immediacy." Audiences can accept significant discrepancies between what they see and hear if some of the music's core auditory elements are represented visually.

Putting It All Together

After explaining its different components, we arrive at the complete model of the musicking quadrant, as sketched in figure 1.8. This model is, obviously, a coarse simplification and generalization of the complex reality of musicking. Nevertheless, it attempts to separate some quite distinct processes, roles, and products of today's music world. The musicking quadrant can also be used to understand more about how the traditional roles are changing. Today, machines enter all parts of the musicking

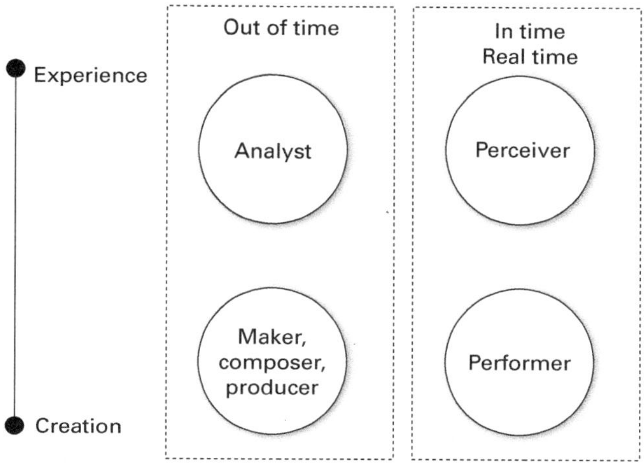

Figure 1.8
The musicking quadrant, identifying four distinct types of musical engagement that can be placed along two axes: creation versus experience and in-time and out-of-time processes.

quadrant. This is exciting, but it may also lead to some challenges, which we will discuss in later chapters.

The different roles in the musicking quadrant should not be seen as mutually exclusive. We have noticed an increasing level of specialization and professionalization in the music world over the last centuries. At the same time, there are also many examples of one person taking on several roles. One person could go and make a flute, come up with some idea for a song, and play it. This would then include the roles of instrument maker, composer, and performer. Another example of such a blending of roles is that of an "open-mic" session. Many of the people present here would partake in the musicking, both in the roles of performer and perceiver.

When it comes to the analyst role, I believe this should extend beyond the work by musicologists. Any type of musical reflection or thinking back at a performance would fall into this category in my model. Both the analyst and the perceiver are on the experiencing side of the musicking spectrum. The main difference is that the perceiver's experience is happening in time (and real time) with respect to the performance, while the analyst's activity is happening out of time (in either real time or non–real time). These may blend, however. An analyst may have been a perceiver at a concert and later relisten to the performance from a recording. The latter would be an example of real-time listening, but out of time concerning the performance. The analyst can then do a non–real-time (and out-of-time) performance analysis using different music analytical tools.

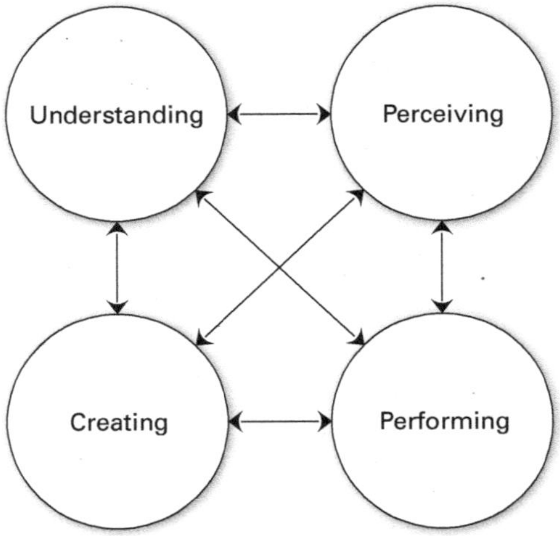

Figure 1.9
Instead of talking about actors, we may think of the musicking quadrant as identifying four distinct types of musical engagement: creating, performing, perceiving, and understanding music.

As these examples show, it is easy to find combinations of the different roles of musicking in the world of traditional (acoustic) music. New roles appear with new technologies. The music producer is one such example. At first, this role was only concerned with physically recording music onto a fixed medium, in time and in real time. Using new technologies, today's music producers largely overlap with composers and performers. In some cases, music producers also use advanced music analysis tools as part of their production chains. One such example is the use of pitch-tracking software to align recorded sound with scores.

The advent of mobile phone apps for "active listening" makes it possible for perceivers to take on a performer role. The developer of those apps—the "digital luthier" to borrow a term from Jordà (2005)—is not any longer only an instrument builder, but could be seen as adding musical content like a composer. We will look more at such cases later and use the musicking quadrant to understand and define the roles. For those discussions, however, it may be more relevant to focus on processes than actors: the functions of creating, performing, perceiving, and understanding music (figure 1.9). In later chapters, we will look more closely at how these processes interact.

2

Music as an Embodied Process

"It doesn't make sense to sit there and listen to this music; you need to get up and dance," the woman said as she gently pushed my fellow students onto the dance floor. She was standing in a "bunad," a traditional Norwegian folk costume, in front of my group of first-year music students. We had been sent to the idyllic village of Voss in western Norway for a one-week intensive course in folk music and dance. We all thought that we would spend most of the time playing and listening to music. As it turned out, the days were split equally between playing and dancing. We quickly discovered the embodied nature of the various Norwegian folk music traditions. You need to dance, to understand the asymmetrical triple meter of Telespringar tunes. Norwegian folk music is no more special than other folk musics. In most cultures there are close relationships between music and dance, what Haugen (2016) has called "music–dance" to emphasize the connectedness. Some languages even only have one term to describe the shared activity of playing, dancing, and singing, such as the word "ngoma" in pro-Bantu linguistic-cultural groups (Bjørkvold 1992; Sarath et al. 2016). But the body is essential for the musical experience also when one is not dancing.

The Body In Music

Even though Small (1998) approached musicking from a sociocultural perspective, many of his reflections resonate well with an embodied approach to understanding the experience of music. This can be seen in his analysis of the strict behavioral "laws" imposed on both performers and perceivers in the classical concert hall. In such a setting, the audience is expected to sit quietly, with as little motion and emotional expression as possible. Even with such rules suppressing the body and its motion, Western art music concerts are also embodied. There are numerous references to body motion in sheet music, such as metaphors like "staccato," "ritardando," and "lento." Such motion qualities embedded in the scores can probably be traced back to the composer's imagined sensations of bodily activity. These imagined motion qualities may (or may not) be interpreted in the performed music through the performer's actions and may, in turn, also lead to physical motion in the perceivers. So even in score-based music, there are close connections between body and mind. Many other musics—and notably different types of dance music—are created with motion as a core element.

Given the apparent importance of the body for both music performance and perception, it is strange that it was only relatively recently that the study of the body's role in music took off. In twentieth-century musicology, relatively little attention was devoted to the topic, even though several people argued for its inclusion, such as Clynes (1982, vii):

> There is much music in our lives—yet we know little about its function. . . . [T]he coming years are likely to see much progress towards providing answers and raising new questions. These questions are different from those music theorists have asked themselves: they deal not with the structure of a musical *score* . . . but with music in the flesh: music not outside of man to be looked at from written symbols, but music-man as a living entity or system.

Clynes's reflection was a response to a more general trend in psychology and cognitive science, which has later led to the development of what is nowadays often called embodied cognition.

Embodied Cognition

An embodied approach to cognition may be seen as an opposition to a traditional cognitive view of the body and mind as separate entities (Shapiro 2019). Standard cognitive theories are based on the idea of the brain as a "device" that passively receives inputs from the world, processes them, and outputs responses. On the other hand, embodied cognition takes the body and its perceptual and motor capacities as having a fundamental role in understanding how we experience the world. This not only relates to how we interact with our surroundings. The interaction is also the basis for creating meaning in the world. From an embodied perspective, meaning is not based on abstract concepts and ideas, but rather something that we create through our bodily interactions in the world (Dourish 2001).

One of the core elements of embodied cognition is the close relationship between action and perception: the "action–perception loop." Hurley (1998) refers to cognition as a "sandwich" of action and perception

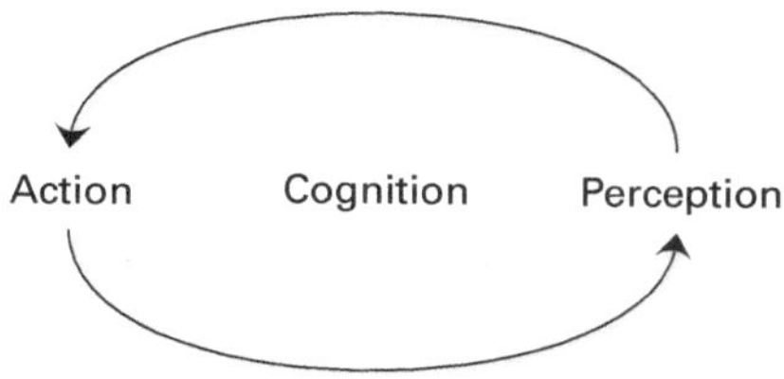

Figure 2.1
A sketch of what Hurley (1998) called the cognition "sandwich."

(figure 2.1). The idea here is that perception is not based on the passive reception of external stimuli; perception guides one's actions. Or, in the words of Noë (2004, 1), "Perception is not something that happens to us, or in us. It is something we do."

Noë (2004) exemplifies how such an action–perception loop works in real life by referring to a blind person moving around a room with a stick. The active tapping with the stick helps discover the space by listening to the sonic responses from the tapped objects and the room's acoustic properties. Blind people often develop a sharp auditory sense, but the same principles are also used by seeing people. If I move my head, I will see something different than if I look straight ahead. If I walk to a different location, I will hear something else than in my current location. As such, the action–perception loop is bidirectional; perception predicts action, and action predicts perception (Schütz-Bosbach and Prinz 2007).

In his ecological psychology, Gibson (1979, 222) stresses the importance of a person's relationship with the environment:

> One sees the environment not just with the eyes but with the eyes in the head on the shoulders of a body that gets about.

The environment should here be understood as anything external to the person in question. It includes the physical space one is within but also objects and other people. Figure 2.2 illustrates this by sketching the constant interaction between the body, the mind, and the environment. There are still many holes in our understanding of how such interactions work in the brain. Current research tries to verify various proposed theories. One of these is the "dynamic systems theory," which suggests that we learn to live in the world through spontaneous self-organization in complex interactions between the body and the environment (Lorenz 1963). These interactions occur at different spatial scales and at multiple time scales. According to the "theory of neuronal group selection," the brain starts early in life with a primary neuronal "repertoire" (Edelman 1987).

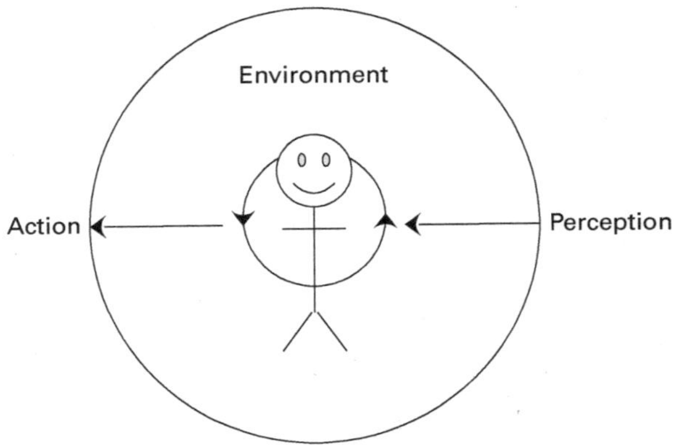

Figure 2.2
A sketch of an "internal" loop between mind and body and an "external" loop between body and environment.

New neuronal combinations are constantly formed, and the combinations that provide the best "solutions" are strengthened and further developed.

The three theories mentioned above—the action–perception loop, the dynamic systems theory, and the theory of neuronal group selection—can be seen as complementary. The action–perception theory describes interactions between an individual and the environment. The dynamic systems theory explains how movement patterns change depending on context. The theory of neuronal group selection describes how the neuronal system adapts to changing stimuli. As such, all the theories acknowledge that the world is complex and that our interaction with the world is based on tackling this complexity by continuously adapting our behavior. The theories are also based on the idea that there can be multiple ways of handling a problem. That is why we can perform actions, say moving a chair, in many different ways. Such a flexible approach to living is learned through childhood, and the interactions are developed as we get older and acquire more skills.

Embodied Music Cognition

In the book *Ways of Listening*, Clarke (2005) suggests that an embodied approach to music perception should be based on "ecological listening." This involves taking our everyday listening and our auditory system's capacities as the point of departure for understanding meaning in music. A large part of our continuous listening is concerned with discriminating between different sound events, what Bregman (1994) calls "auditory scene analysis." Through evolution, we have adjusted our hearing to what is presumably crucial for our survival. Clarke argues this evolutionary-developed auditory system is the basis for composing, performing, and perceiving music. Thus, our perception of musical sound should be studied regarding our auditory system's capacities and limitations.

In his book, Clarke explains how we approach new sonic/musical material through three reasoning steps. The first level is about identifying "what sounds are the sounds *of*, and what to do about them" (Clarke 2005, 3). Such "sonic causality" can be seen as identifying which objects are involved in the sound production. The second level of listening is connected to the meaning gained when you understand what a sound is the sound of. The third level is related to how the sounds layer and interact with each other. We usually explore many sounds simultaneously, and what could be considered "foreground" sounds are also being merged into sounds of the "background."

What differentiates the ecological listening theory of Clarke (2005) from more traditional music psychological approaches is the inclusion of culture in the discussion. This is illustrated in an analysis of Jimi Hendrix's performance of "The Star-Spangled Banner" at Woodstock in 1969. The meaning(s) attached to this performance—and its impact—needs to be understood as a complex interplay between the cultural contexts that the performance happened within and the musical features. Clarke convincingly shows how an interpretative musicological analysis can be combined with a cognitive framework.

One limitation of Clarke's approach in *Ways of Listening*, which he acknowledges early on in the book, is the primary focus on sound. While his argument is certainly based on the idea that listening to sound is a multimodal phenomenon, he does not develop his argument into a fully embodied approach. This is what Leman (2008) tries to do in his book *Embodied music cognition and mediation technology*. Here, he sets out to create a complete model of musical communication, as sketched in figure 2.3. Musical intentionality is at the model's core, and Leman argues that a performer's intentions are only indirectly communicated. Direct communication is based on auditory and visual cues. As such, the performer's actions are overt representations of covert intentionality.

In Leman's model, the action–perception loop is represented through an internal loop between "corporeal" articulations and imitations. What happens in the body is mirrored in the brain. Similarly, what happens in the brain is the starting point for further actions. This is not limited

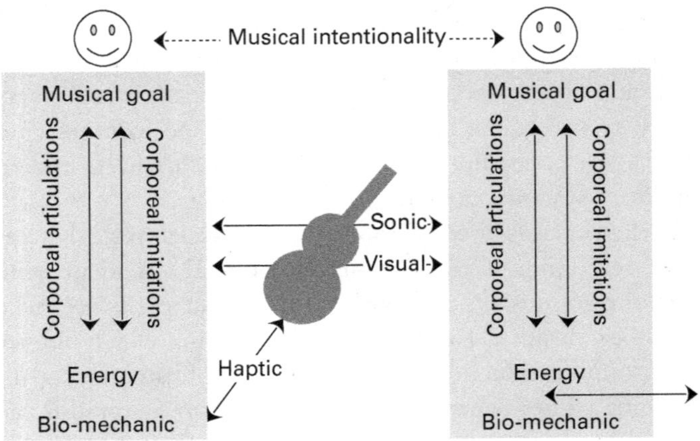

Figure 2.3
A model of embodied music cognition, based on (Leman 2008, 160) (see text for explanation).

to performers. Perceivers also have an ongoing internal loop between mind and body. This may be why we have an urge to move when listening to music or get physical sensations when seeing other people play an instrument.

Leman describes the performer's multisensory interaction with the instrument, including auditory and visual feedback and haptic stimuli. The perceiver only sees/hears the auditory and visual stimuli but does not get any haptic stimuli from the instrument. However, based on previous haptic experiences and an internal loop between mind and body, one can speculate that a perceiver gets a sense of the haptic feedback by only listening/seeing a performance. For example, when I watch someone play the guitar, I can "feel" how it is to touch the strings. This type of sensation I can get by only imagining someone playing the guitar. It is precisely these types of complex loops between the minds and bodies of both performers and perceivers that Leman's model tries to explain.

Motor-Mimesis

When it comes to our mental representations of music, or what could be called "musical imagery," Godøy (2003, 318) has suggested that these may be based on "motor-mimetic" principles:

> Motor-mimesis translates from musical sound to visual images by a simulation of sound-producing actions, both of singular sounds and of more complex musical phrases and textures, forming motor programs that re-code and help store musical sound in our minds.

This builds on the "motor theory of perception," which was proposed in linguistics to explain the close connections between action and sound in speech perception (Liberman and Mattingly 1985). Here the claim is that language perception and acquisition are based on learning the "articulatory gestures" of the human vocal apparatus. Such articulatory gestures can be seen as the basic "atoms" out of which phonological structures are formed (Browman and Goldstein 1989). The motor theory of perception further postulates that the phonological, lexical, and syntactic structures are based on neural systems that have evolved for general motor control. This implies that the perceptual mechanisms of the vocal apparatus were in place long before language entered the stage and that speech and language adapted to phylogenetically older structures rather than the other way around (Lindblom 1991).

The motor theory of perception received renewed interest after Gallese et al. (1996) found "mirror neurons" in monkeys' brains. These experiments showed that a particular set of neurons were activated both when a monkey performed a goal-directed action and when observing another monkey performing a similar action. These and later experimental results support the idea that we continually simulate the actions we see. Such a mental "replay" of actions probably helps to learn. It may also explain our ability to make predictions. Wilson and Knoblich (2005) argue that we continuously simulate our actions in parallel to the physical actions carried out. Such imitative motor activation feeds back into the perceptual processing and allows us to adjust our actions accordingly.

The early experiments on mirror neurons were of goal-directed actions observed in monkeys. Later, Rizzolatti and Arbib (1998) carried out positron emission tomography (PET) studies of humans and found that motor areas in the brain were activated during speech perception. Such an "auditory-motor interaction" has also been shown in functional magnetic resonance imaging (fMRI) scans of humans (Hickok et al. 2003). Here similar neuronal activity was found when subjects only listened to speech or music and covertly hummed the auditory stimuli.

Another set of studies has shown the existence of "audiovisual mirror neurons" in the brain. This was found in experiments on monkeys, where the same neuronal response was measured whether the monkey could both see and hear a sound-producing action or only hear the sound of the same type of action (Kohler et al. 2002). There have also been studies of audiovisual mirror neurons with more complex sounds. For example, Haueisen and Knösche (2001) used magnetoencephalography (MEG) to show pianists' involuntary motor activity while listening to well-known piano music. Lahav et al. (2007) reported that similar motor activity might occur in nonmusicians. They carried out an experiment during which nonmusicians practiced playing a melody for a couple of weeks (learned by ear). After this practice, fMRI scans showed that motor areas in the brain were active when the subject listened to the same melody as had been practiced and when the person was not moving. This activation would only occur when listening to the same melody the person had practiced; variations on the melody resulted in little or no activation. This can be thought of as an "action-listening" process during which we continuously simulate the sound-producing actions related to the sounds we hear.

How does the discovery of mirror neurons, and the growing body of empirical evidence supporting the motor theory of perception, help develop the field of embodied music cognition? In his "mimetic hypothesis," Cox (2016) proposes twenty principles for explaining musical experiences. His approach focuses on the perceiver and does not include the feedback loop between the performer and perceiver that we find in the model of Leman (2008). On the other hand, Cox digs deeper into the idea that perceivers experience music through embodied imitation of what he calls the "bodily motions and exertions" involved in producing music. He also makes a distinction between "overt" and "covert" mimetic behavior. Overt behavior, such as moving the arm, he calls "mimetic motor action" (MMA). Covert behavior is called "mimetic motor imagery" (MMI) and is the experience of muscle tension without motion. Of his twenty principles (Cox 2016, 13), some are of particular interest for my further argument:

- Principle 1: Sounds are produced by physical events; sounds indicate (signify) the physicality of their source.
- Principle 2: Many or most musical sounds are evidence of the human motor actions that produce them.
- Principle 3: Humans understand other entities (animate or not, human or not) and events in their environment in part via mimetic behavior (MMI and MMA).
- Principle 5: Mimetic comprehension is based on visual, auditory, and/or tactile information.
- Principle 11: Any and all acoustic features can or will be mimetically represented.
- Principle 12: Different kinds of music "invite" (motivate) different kinds of mimetic engagement, contributing to the feel (quale) of music.

The three first principles are similar to my thinking about action–sound couplings that we will get to in chapter 7. In the following, we will look more at two concepts from principles 5 and 12: multimodality and affordance.

Multimodality

A fundamental aspect of embodied (music) cognition is that our perception is inherently multimodal. This means that all our senses and modalities (as summarized in table 2.1) are present when experiencing the world. That is why I prefer the term "perceiver" over "listener." Many people say that they think about listening as a multimodal activity. However, I like to use a word that reminds me that all the senses are at work. Then it is also possible to reserve the term "listening" to the act of auditory perception.

One problem with using "perceiver" as a general term is that it may sound like the focus is only on the sensory input and not on the cognitive processes. One could argue that a word like "experiencer" could be better, but that may be a bit too broad again. For now, I have therefore decided to use "perceiver" as my primary term.

When I first started using "perceiver" instead of "listener," it felt awkward, and I got many questions. After using it for some years, it feels normal. It has been so internalized that I rarely think about it any longer. Someone once asked me if I did not mean "receiver" instead of "perceiver," pointing out that the "reception of music" has been used in some musicology literature. My answer was that these are entirely different terms. I think that a receiver is the opposite of a perceiver. A model based on having a sender and a receiver signals unidirectional and passive

Table 2.1
The human senses and their corresponding sensory organs and modalities (Schomaker 1995, 5).

Sensory perception	Sensory organ	Modality
Sight	Eyes	Visual
Hearing	Ears	Auditory
Touch	Skin	Tactile
Smell	Nose	Olfactory
Taste	Tongue	Gustatory
Balance	Organ of equilibrium	Vestibular

communication. The embodied music cognition paradigm, on the other hand, is built around the idea of a continuous feedback loop between performer and perceiver.

The senses and modalities work in parallel, and they also mutually influence each other. Let us set up a thought experiment of the interdependence and inseparability of our modalities. Think about walking down the street, while suddenly you hear the sound of a guitar being played. Most likely, you will turn toward where the sound is coming from and see who is playing. Already at this point, three modalities have been at work: the auditory, the visual, and the vestibular. First, your hearing's omnidirectional nature recognizes the sound. Then you turn so that the directionally limited sight can see the location of the sound source. Here the vestibular modality is vital for keeping you in balance and informing you about the orientation and motion of your head and body. This seemingly simple example of an everyday musical experience shows some of the richness of our multimodal capacities.

An exciting aspect of our multimodal perception is that we rarely pay attention to which modality is at work, and it often does not matter. For example, think about how you recognize other people. This can be through hearing their voice, seeing their way of moving, smelling their perfume, or a combination of these. We rarely think about which senses are at work, even though we may choose to focus on one particular sense if we want to. The senses work together most of the time, helping us move around in a complex world with a continuous stream of perceptual input.

Our multimodal capacity may also give rise to "crossmodal" behavior, such as in synaesthesia, sensory substitution, and crossmodal perception. The McGurk effect is a famous example of the latter. McGurk and MacDonald (1976) carried out an experiment where the subjects heard the spoken word "Ba" while seeing the video of a mouth saying "Ga." However, the subjects did not perceive any of these two words but saw/heard "Da," a combination of the two stimuli. There have been attempts to find similar types of audiovisual integration in music. In a clarinet performance study, Vines et al. (2006) showed how visual information enhanced the sense of phrasing and the anticipation of changes in the emotional content. This was done by presenting video recordings to the subjects in which the audio had been substituted. Thompson et al. (2005) reported a similar visual impact on the perception of music based on case studies of recordings of popular artists.

Given the close connection between perception and action, there are numerous examples of using one modality in perception and another in action. For instance, in conversation, one may answer a spoken question with a hand gesture, thus combining the auditory and visual channels in the communication. Such crossmodal communication can even be seen in infants (Trevarthen 1999), suggesting that this is a fundamental capacity of the human cognitive system.

Multimodal perception and crossmodal interaction are central in music performance and perception. Hearing the sound is crucial, but also the haptic and visual feedback of the instrument are essential for the performer. The vestibular modality is critical for percussionists moving around while playing. Wind instrumentalists rely on their gustatory capacities to wet the reed. Perceivers rely on both hearing and seeing during a concert. But also the smell of a concert hall, the tactile feel of sitting in a chair, or the sense of vestibular activity when dancing may all play a role in how we experience music. Even when people's engagement with music is based on listening through headphones, music continues to be a multimodal experience based on the body's presence and motion (Zelechowska et al. 2020).

Affordance

One fascinating aspect of our cognitive capacities is understanding how we can interact with the objects we encounter. For example, if we encounter a sharp object, like a knife, we know it can be used to cut something. Alternatively, we may easily see that a wooden object with four legs and a base is something that we can sit on. We would probably immediately categorize such an object as a "chair," even though we may never have seen this particular chair before. The chair could be of many different sizes, have different shapes, have a different number of legs, and be made in a large variety of materials, fabrics, and colors. Given all these differences, we would still call it a chair. Even if there is a table in front of it, we can quickly separate the chair from the visual scene. This may seem trivial, but the development of computer-based "visual scene analysis" to do the same has exposed such a task's complexities.

One explanation for why humans are so good at performing visual scene analysis may be our capacity to identify "motor programs" associated with objects (Rosch et al. 1976). This means that we recognize a chair because of its function; we can sit on it. Thus the motor program of "sitting" is the essential element when we categorize an object as a chair. This is not unique for humans. Jakob Von Uexküll gave already in 1920 an example of how an oak tree could be used in different ways (de Haan et al. 2013): providing a rabbit the opportunity to dig a hole between its roots, providing a woodworm with food, and giving a person a place for shelter.

As part of his ecological psychology, Gibson (1977, 127) introduced the term "affordance" to describe our ability to understand the action possibilities of objects:

> The affordances of the environment are what it offers the animal, what it provides or furnishes, either for good or ill. The verb to afford is found in the dictionary, the noun affordance is not. I have made it up. I mean by it something that refers to both the environment and the animal in a way that no existing term does. It implies the complementarity of the animal and the environment.

To continue with the chair as an example, we may say that it affords sitting on. However, it may also afford being a table, being thrown at someone, or being used as a percussion instrument. As such, a chair—and any other object—may have multiple affordances. The affordances depend on a person's experience interacting with the object and the cultural context within which the object is located and used. For Gibson, affordances exist as a relationship between the properties of the environment and the possibilities of the user. For example, a cup that is graspable for a grown-up may not be graspable for a child. Thus the "appropriation" of the affordances differ.

It may be argued that some affordances are more fundamental than others, but most affordances are probably learned by living. We continuously expand our knowledge of affordances through daily interaction with objects in the world. We can also easily come up with new usage of objects. For example, even though we have learned that a chair is made for sitting, we may easily create a long list of other types of uses that the chair affords. Thus, imagination and creativity are also essential aspects of our embodied cognitive system.

The affordance concept has become popular in design theory, notably after Norman (1988) used it in his popular book *The Design of Everyday Things*. His definition of affordance came from a design perspective, focusing on creating useful objects, and was, therefore, more limited than Gibson's more general definition. Realizing that his use of the term had caused some confusion, Norman spent several pages explaining his (mis)interpretation of Gibson's term in the revised version of his book, *The Design of Everyday Things: Revised and Expanded Edition* (Norman 2013). Norman's use of the term is mainly concerned with a designed object's visual appearance. I will use the term when describing musical instruments. Then it is necessary to expand its notion to also include an object's sonic properties.

Sonic Affordances

Most objects produce some kind of sound when they are acted upon. For example, moving a chair will typically result in the sound of its legs scratching the floor. This sound is part of what I will refer to as a chair's "sonic affordances," the possible sonic outcomes based on the properties of an object. Let us explore this idea by thinking about a glass falling toward the floor. If the falling glass is made of plastic instead of glass, it will probably bounce off with a plastic-like sound once it hits the floor. This would be a dramatically different outcome than if the glass would break into pieces. Thus, if we thought the glass was made of glass, we would probably be surprised by a plastic-like sound, but it would still be possible if we were unsure of its construction. However, if we heard a baby's scream when the glass hit the floor, it would be beyond the laws of nature, and we would assume that the scream was coming from elsewhere. This is simply because a baby's scream is not within the span of sonic qualities afforded by the objects and actions involved. I will call these multiple sonic affordances the "action–sound palette" of an object. As will be discussed more in chapter 7, I believe that the action–sound palette of an instrument is vital for our perception of its action–sound couplings or mappings.

My use of the term "action–sound palette" is somehow similar to the idea of a "landscape" of affordances proposed by de Haan et al. (2013). They have a psychiatry perspective, in which three dimensions of such a landscape are used to understand their clients. The first dimension is the "width," which refers to the broadness of the scope of affordances that one perceives. This relates to having different options for action. The "depth" dimension refers to the temporal aspect. We perceive affordances here and now, but we are also constantly aware of future possibilities for action. This is an integral part of anticipating and predicting what will come and what we can do. The third dimension is the "height," which refers to the importance of the affordances that one is responsive to. In their thinking, the landscape of affordances is highly subjective. Although this landscape metaphor was developed for a different purpose than music, I find it helpful to explain affordances' complex and subjective nature in general.

Another extension of Gibson's thinking about affordance is proposed by Withagen et al. (2012). They suggest that affordances not only are action possibilities, but also "invite" behavior. Their approach is coming from industrial design, which is largely based on designing and creating objects that make certain behaviors more likely to occur. In a discussion of affordances and agency, they list four factors that are of importance. The first is that of the "action capabilities" of the agent. Together with the environmental properties, such capabilities determine what actions can be performed and what actions are invited. That is, action capabilities determine how affordances can be appropriated. The second feature is that some affordances are more critical than others. If the fire alarm goes off, you will flee the building instead of playing the piano. The third feature is related to the importance of culture. One's cultural background gives rise to particular responsiveness to specific affordances in the environment. The fourth is based on individual differences. Even though there are many similarities between people from the same culture, there are also many differences. For example, chocolate may afford eating for many, but not everyone.

In a study of playground designs, Withagen and Caljouw (2017) discuss how abstract playground elements afford more types and more varied play than single-purpose elements. It has also been found that children find it more exciting when playground elements—such as jumping stones—are nonstandardized, as this appears to afford more play (Sporrel et al. 2017). The same principles may also be found in musical instruments. Hunt et al. (2003) found that users preferred more complex and unpredictable mappings over simpler ones in their instrument designs. However, the performer's skill level may be a factor. Rietveld and Kiverstein (2014) suggest that the affordances that the environment offers to an animal are dependent on the skills the animal possesses. From a musical perspective, a tablespoon may offer more affordances to a professional percussionist than a toddler. However, the toddler would probably be less socioculturally constrained than an adult.

In a discussion of "musical affordances," Menin and Schiavio (2012, 211) criticize previous attempts at fitting Gibson's affordance definition to music. They propose a motor-based interpretation of musical affordances:

> The situatedness of musical affordances, indeed, as it is rooted in action-understanding processes rather than in mental forms of the transmission of knowledge, cannot be decontextualized from the subjective goals. . . . Through the manual exploration of objects, infants develop their motor acts and familiarize themselves with musical structures such as repetition and variation. The knowledge acquired from those discoveries will make the children able to construct a musical context as well as a basic vocabulary of musical-directed acts.

I find the concept of affordance particularly useful when discussing musical instruments because it helps explain the relationships between action and sound. For example, a string-based instrument affords plucking or bowing, which again will result in plucking-like or bowing-like sounds. You would never expect blowing-like sounds from a guitar nor plucking-like sounds from a trumpet. Thus, our understanding of musical sounds is mostly based on the construction of instruments and the types of performance actions they afford. This is usually straightforward for acoustic instruments but more challenging for electro-acoustic instruments. For example, keyboard-based synthesizers have an interface that affords piano-like actions. Acoustic piano performance is based on impulsive actions. However, a keyboard can be used to control any type of sound model. Playing sustained sounds—such as the sounds of string or wind instruments—violates a basic affordance of the interface. It also breaks with an expectation that a violin-like instrument should play a violin-like sound. This may make the instrument feel unnatural and inexpressive to both the performer and the perceiver.

3

Musical Instruments

"Is it a real saw?" Many people comment on the saw I keep on a shelf in my office. And, yes, it is a real saw, although it is branded as a "musical saw" and sold under the name Sandviken Stradivarius (figure 3.1). However, it does not differ much from the ordinary, well-used, crosscut hand saw I use for cutting up wood for the fireplace at home. The main difference is that the musical saw has a logo of a saw musician on the side. Does that make it into a musical instrument? No, if we think that saws were originally designed for cutting wood. However, the Sandviken Stradivarius is designed, manufactured, and branded as a musical instrument. I occasionally play it in my office, so my exemplar is used for musicking. The same is true for my other saw. I have played on that one, too, even though it was not made for performance. My two saws were manufactured with different intentions, but if one disregards the logo, they look alike. Both can produce sound and have music-making potential. They can also both be used to cut wood, although I am pretty sure that I would not reach for the Stradivarius first.

What Is a Musical Instrument?

An oboe is a musical instrument. So are charangos, clarinets, and castanets. These are all physical devices made with the intent of performing music. What about a spoon, a bucket, or a chair? Are they also musical instruments? From a design perspective, they are not since they were not originally made for musical activities. However, from a user's perspective, a spoon may be turned into a musical instrument if used in a performance. I find these dichotomies between design intentions and usage fascinating.

For a long time, music research has focused on works (Goehr 1992) and the composition and performance of such works (Johnson 1997). The

Figure 3.1
The Sandviken Stradivarius (bottom) is a saw made for music performance. How does it differ from an ordinary crosscut hand saw (top)?

works have traditionally been studied as fixed products, either in the form of scores (compositions), recordings (of performances), or productions (of pieces/tunes). Much less attention has been devoted to the role of musical instruments, how they are made, how they look, how they feel, and how they sound. This is the case even though "organology"—the academic study of instruments—has been a well-established discipline for more than a century. Even so, when I ask incoming bachelor's students about what they know about organology, they stare at me with blank eyes.

"Everyone" agrees that instruments are essential for musicking. Still, instruments receive relatively little attention in musicological discourse. There are some notable exceptions, including investigations into the "social life" of instruments (Bates 2012), thoughts on instruments as "creative prostheses" (De Souza 2017), and reflections on twenty-first-century instruments (Bovermann et al. 2017). Yet, often the importance of instruments is forgotten. In my opinion, a better understanding of musical instruments is key to understanding how musicking at large is in transition. From here on, we will shift our attention to musical instruments and look at their construction and function in both acoustic and electro-acoustic musicking.

Instrument Identity

The musical saw represents a borderline case between design intention, usage, and cultural connotation. This example also shows that defining an instrument is not as easy as one may think. As we shall see later, it becomes more complicated when considering electro-acoustic instruments. But is this a problem? Does it matter how we define a musical instrument? Kvifte

(2008, 55) cautions against trying to define what an instrument should be, worrying that "if a traditional and relatively precise definition of "instrument" excludes large areas of contemporary musical practice from our field of study, we might be better off with less precise alternatives."

In my experience, instrument definitions are more important than we like to acknowledge. For example, in higher music education, the norm is still to play "an instrument." I am constantly reminded about this when I participate in our department's welcome lunch for new music students each fall semester. Listening to the students introducing themselves is fascinating. They usually start by stating their name and the school they attended before coming to university. Almost without exception, the third thing they mention is what instrument they play. Their instrument defines who they are as aspiring professionals. Some students say they play "the piano." However, this often leads to follow-up questions about genre. Playing "classical piano" or "jazz piano" is not the same. Their sound maker—the physical piano—may be the same, but their use of it differs. It helps to add the genre to the naming of their instrument to explain their musical identity. As such, the genre becomes a subcategory of the naming of the physical instrument. This may not be as strange as it sounds. In many cases, the performance technique differs between genres. If you say that you play "contemporary piano," one may think that you spend a considerable amount of time fiddling inside the piano, touching the strings, adding objects, or developing extended techniques. This is different from playing "classical piano."

For piano, it is common to add the genre as a prefix to the instrument's name. There are also examples of how the naming of the instrument itself is dependent on the genre. Kvifte (1989) discusses the distinction between the instrument(s) "fiddle" and "violin." In most cases, these terms refer to the same type of instrument. An exception is the Norwegian Hardanger fiddle, with four resonating strings below the normal strings, although this instrument would probably be called "Hardanger fiddle" and not only "fiddle." When it comes to normal fiddles and violins, they are the same physical instrument. However, they are often played differently. Most violinists would rest their instrument toward the chin, while fiddlers may rest it on their arm. Some Indian classical musicians sit cross-legged on the floor, resting the violin's head at the sole of their feet (Magnusson 2021). Not only do the genres lead to different naming of the instruments, but also the naming of the performers. A "fiddler" is not the same as a "violinist." This is not just a question about terminology; it also defines a musician's training and career.

There are also examples of how the naming of an instrument may cover different types of physical instruments, such as when people say "I play keys." In most cases, this would indicate that the person plays one or more types of keyboard-based instruments. This often includes both acoustic piano and various kinds of analog and digital keyboards. Most likely, "keys" would also indicate that the person plays popular music or jazz. I have still to come across a classical performer stating that they play "keys."

What instrument you play matters. After all, professional identities are built around instruments. Being "a fiddler," "a pianist," or "a DJ" defines what you do and who you are in the music world. Conservatories and universities recruit students based on instrument type and so do most traditional orchestras ("Searching for a violinist...") and bands ("Looking for a bass player..."). Also, many nonmusicians have a clear idea about what the different terms mean. When I tell people that I work with music, people always ask about my instrument. It is difficult to answer such a question. I could be honest and say that I design, build, and play various body-based electro-acoustic instruments. However, if I do not want to get into lengthy explanations, I say that "I play the piano." That is a straightforward answer, and it is not entirely false given that the piano is my "mother instrument," so to speak. The normal follow-up reply, then, is something like "I used to play the piano as a child." This is the case even though the person may not have touched the instrument in decades. Still, a person's instrumental childhood experience is something they carry on as part of their identity.

As electronics have become more popular, it is nowadays quite common to hear students say, "I play laptop" or "my DAW is my instrument." I find such statements interesting because they indicate the shift that this book is all about. We will return to such cases later. In the rest of this chapter, we will primarily focus on traditional acoustic instruments.

Instrument Definitions

It is clear that the naming of instruments matters, but we have not yet come closer to a proper definition of an instrument. As the discussion so far has revealed, coming up with a proper definition is not easy. The Grove Music dictionary ultimately resigns in their article on "instrument" (Brown 2001):

> Like many other musical terms, however, the word meant various things at various times, and it was not always used consistently.

Their entry on "musical instrument" is better at explaining the historical notion of instruments and their limitations (Libin 2018):

> Conventionally the term refers to implements specially designed for producing sound, but this definition is inadequate because unaltered natural objects as well as utensils meant for other tasks (nowadays including electronic communication devices) have been put to musical use since prehistoric times.

This resonates well with my thinking, although the terms "implements" and "utensils" do not sound particularly musical. The final definition of an instrument is more focused, though, and is the one that I will continue to use in this book (Libin 2018):

> Vehicle for exploring and expressing musical ideas and feelings through sound.

There are several reasons why I like this instrument definition. One is that sound is at the core of the definition. On a side note here, one could always question whether you need sound to call something music. For example, there have been several experiments with creating "visual music" (Evans 2005; McDonnell 2020). We will leave such borderline cases aside for now, though, and assume that sound—including the sound of silence (Cage 1961)—plays a significant role in musicking at large.

I also like Libin's use of vehicle instead of, say, tool or device in the definition. This opens for including the human body and the human voice as instruments. The definition further opens for both exploration and expression, indicating an active and embodied thinking about musicking. The same can be said about including both ideas and feelings, acknowledging that musick covers both structures and emotions.

In a linguistic study of musical instrumentality, Cance (2017) concludes that the term "instrument" refers to a device and its interaction with the users. I will use such an application-focused definition of musical instruments as a point of departure. Generally speaking, we may say that an instrument can be a mediator between action and sound (Bielawski 1979; Kvifte 1989), as sketched in figure 3.2. An instrument can be a physical device, but also the human body, a virtual device, or even just an imaginary idea. We will get back to all of these later.

Figure 3.2
A musical instrument can be seen as a mediator between action and sound.

When I emphasize sound in the definition of a musical instrument, this does not necessarily mean "musical" sound. In the same way that a spoon may be used to create musical sound, a musical instrument may be used to make nonmusical sound. Accidentally dropping a guitar on the floor will produce sound, but it would usually not be considered musical. With our current definition of an instrument, any device, from spoon to computer, may be considered a musical instrument if used to make music. As we shall see later, the musical content and context may even be integrated into the instrument itself. However, before exploring such complexities, we will rewind a little and consider some traditional ways of thinking about instruments.

The Hornbostel Sachs System

Up until the late nineteenth century, instruments were usually sorted into three categories: strings, wind, and percussion (Lee 2019). Based on work at the instrument museum in Brussels, Mahillon (1880) proposed a division into four types, based on the sound production of the instruments: autophones, membranes, wind, and chords. This was further developed into what has become the most well-known musical instrument categorization, often known as the "Hornbostel Sachs" system. This system was first introduced in the book *Systematik der Musikinstrumente* (von Hornbostel and Sachs 1914, 1961). Building on the work by Mahillon, it classifies musical instruments based on their sound-producing elements.

Idiophones: The material itself generates the sound (e.g., xylophone)

Membranophones: A vibrating membrane generates the sound (e.g., drum, kazoo)

Chordophones: A moving string generates the sound (e.g., violin, piano)

Aerophones: A vibrating air column generates the sound (e.g., oboe, clarinet, saxophone)

The Hornbostel Sachs system is categorical, following a tree-like structure with specific subentries. The level of detail is high, albeit limited by its textual descriptors. For example, let us consider the definition of the percussion instrument castanets, which is categorized with code 111.141 (von Hornbostel and Sachs 1961, 14):

1 Idiophones: The substance of the instrument itself, owing to its solidity and elasticity, yields the sounds, without requiring stretched membranes or strings

11 Struck idiophones: The instrument is made to vibrate by being struck upon

111 Idiophones struck directly: The player himself executes the movement of striking; whether by mechanical intermediate devices, beaters, keyboards, or by pulling ropes, etc., is immaterial; it is definitive that the player can apply clearly defined individual strokes and that the instrument itself is equipped for this kind of percussion

111.1 Concussion idiophones or clappers: Two or more complementary sonorous parts are struck against each other

111.14 Concussion vessels or vessel clappers: Even a slight hollow in the surface of a board counts as a vessel

111.141 Castanets: Vessel clappers, either natural or artificially hollowed out

One fascinating—though confusing—part of the Hornbostel Sachs system is that it combines descriptions of the physical construction of an instrument with explanations of how it is performed. In the case of the castanets mentioned above, three of the properties relate to the instrument's construction (1, 111.14, 111.141), while the three others relate to the performance properties of the instrument (11, 111, 111.1). This assumes an agreement on the (only) way an instrument is to be performed. You do not have to be very imaginative to come up with multiple performance techniques for many instruments. This can be problematic from a tree-like classification perspective and impossible when considering extended techniques. Including all such performance techniques would easily break the entire system.

For obvious reasons, the 1914 version of the Hornbostel Sachs system did not include electro-acoustic instruments, but these were added as a fifth category, "electrophones," later (Sachs 1940). As part of establishing Musical Instrument Museums Online, a consortium of European instrument museums revised this fifth category to take into account new developments (MIMO 2011). An overview of the first two levels of the current official version of the Hornbostel Sachs system can be seen in figure 3.3. Here, the fifth category has been expanded to include six types of electrophones.

Other Organologies

Despite its shortcomings, the Hornbostel Sachs system works reasonably well for many traditional instruments. It has met criticism over the years

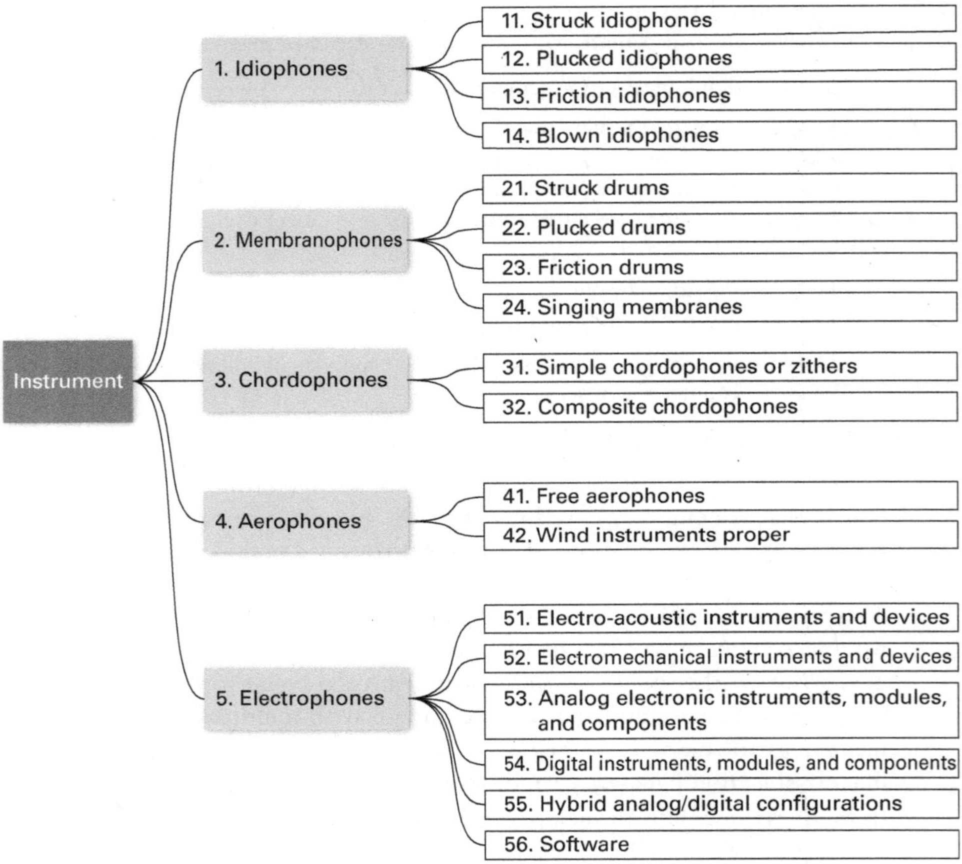

Figure 3.3
The two first layers of the updated Hornbostel Sachs system, based on (MIMO 2011).

(Lee 2019) but continues to be the de facto standard used in instrument catalogs and museums. For such usage, its categorical, tree-like structure helps to categorize different instruments.

There have been several attempts at extending the Hornbostel Sachs system over the years. For example, Olsen (1986) proposed "corpophones" as a category to describe vibrations produced by body parts. Interestingly, he does not include the voice in this category, arguing that musical vocal sounds are songs. "Fictophones" is another proposed category describing imaginary instruments. As Loughridge and Patteson (2013) write in their introduction of the online Museum of Imaginary Musical Instruments:

> Imaginary instruments are a special kind of technological phenomenon. Such instruments never fully make the passage from the imagination into the world. They remain unconsummated objects, indifferent to the chaotic forces at play outside the test-tube of pure conceptuality. Ranging from the physically impossible to the simply impractical, from the "never" to the "not yet," imaginary instruments rattle suggestively at the windowpane separating our comfortable sense of reality from that nebulous space beyond. In the words of Ernst Cassirer, such instruments are "concerned in the final analysis not with what *is*, but with what *could be*."

This way of thinking about imaginary instruments resonates well with my techno-cognitive approach. After all, an imaginary instrument can also lead to musical experiences. Even if there is no external physical representation of any music being played, there are many accounts of vivid experiences based on musical imagination (Grimshaw-Aagaard et al. 2019).

There have also been attempts at developing entirely new organological systems (see overviews in Kvifte 1989; Magnusson 2017). Of these, I find the physics-based organology by Mann (2007) intriguing because he proposes a matrix in which the instrument's sound-producing element is split from its user interface element. This opens for different types of explorations with new instruments and performance techniques. Indeed, for him, the system has also represented a tool for instrument design and musical exploration. Many of the older classification systems were intended for instrument classification for museums and research. I am more concerned with how instruments are used, and I will therefore base my classification on such a division between sound production and sound modification.

In a similar line of thought, Magnusson (2017) suggests that the shared usage of instruments can be considered a pool of possibilities from which the instrument definition can be found. He calls this a "heterarchical approach to digital organology." The aim is to break away from the traditional tree-like classification systems and instead focus on the instrument's material design and technical origins.

I think all of these alternative instrument definitions have their pros and cons. They help to broaden and challenge our thinking about musical instruments. However, a challenge with the more open-ended instrument definitions is that they risk being so "fuzzy" that they are challenging to use in practice. My aim is not to create a system that covers everything. Instead, I want to create a theoretical framework to understand what an instrument is, how it is used, and how it is perceived. This includes understanding more about the differences (and similarities) between acoustic and electro-acoustic instruments.

Embodied Organology

The voice has traditionally been left out of organological classifications. This is not strange if one thinks about instruments as physical objects that can be put on a shelf in a museum. However, the voice should be included if one thinks of an instrument as a vehicle for musical expression.

Kreiman and Sidtis (2011) describe the voice as a "bioacoustical instrument" with a two-stage sound production. The first part is creating a fundamental frequency and related harmonics, which are the basis for the perceived pitch of the vocal sound. The second part is shaping the timbral qualities of the sound through formant frequencies that are resonances of the vocal tract. As such, the voice comprises both sound-producing and sound-modifying parts as any other acoustical instrument. The main difference is that all of the parts are literally embodied within the performer.

The lack of inclusion of the voice in traditional organology also has a cultural dimension. Discussing the changing voice of Maria Callas after her weight loss, Eidsheim (2017, 16) argues for a critical vocal organology:

> Investigating the body, which is burdened by millennia of social practices, is one small step in the process of unsticking the female (and the raced, classed and so on) body and voice from their construction and existence within a social and cultural sphere that can only comprehend them from within its own set of resources and hierarchies. In my assessment, critical vocal organology does so not by avoiding, but by addressing head-on, the sociocultural material dimensions of voice.

Several others have also called for thinking more about the body when describing instruments. Kvifte (2008, 48) criticizes instrument definitions that leave out the human body, commenting that, in traditional organology, "the instrument is what is left behind when the performer is no longer present." He exemplifies this with the trump (also known as Jew's harp, guimbarde, and maultrommel), an instrument that consists of a flexible reed attached to a frame (figure 3.4). This is an instrument that relies on the performer's mouth cavity for sound production and modification. What is then part of the instrument? The mouth of the performer, the lungs, the entire body? If one agrees that the body is part of a trump, what about other instruments with a physical connection between a performer and the instrument? For example, the fingers of a violin player touch the strings and are also essential for music making. Should they, too, be considered part of the instrument? Instead, one may classify instruments after

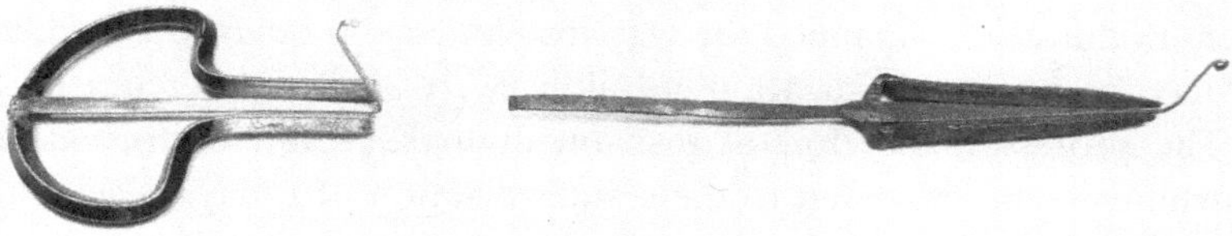

Figure 3.4
Examples of a modern trump (left) and a replica of an old one (right). These instruments require the cavity of the mouth for sound production.

how they are performed. Kvifte (1989) builds on the model by Heyde (1975), separating between "technopomorphic" and "anthropomorphic" elements of instruments. That is, an instrument's function can be split into mechanical and human elements, respectively. Then it is also possible to more directly analyze the interaction between body parts and controllers on the instrument.

A similar line of thought is present in the writings of Paine (2015) on the "techno-somatic dimension." Paine sees this dimension as a meeting point between the physical instrument itself and how the human acts upon that instrument. This techno-somatic dimension contains two parts. The technical relates to instrumental technique and performance practice, while the somatic concerns the bodily experience. This resonates with the techno-cognitive approach I propose in my action–sound theory.

Ethno-Organology

Magnusson (2021) argues for an "ethno-organology" that takes into account how instruments move between cultures. He argues that technology is never independent of societies. Instruments are developed within one culture, but they are also used in others. How does such "instrument migration" affect others?

Ahrens and Zedlacher (1996) discuss the development of wind instruments during the nineteenth century. Instrument makers started equipping woodwinds with a linked key mechanism and brass instruments with valves. These technical developments improved the intonation, facilitated playing, and increased the pitch range of the instruments. However, these improvements came at the cost of a larger action–sound separation. On a silver flute, the fingers are no longer in direct control with the sound hole.

A vibrato can no longer be created by simply moving the fingers over the sound holes. Instead, a vibrato needs to be produced by the lips. Many such modifications changed the way instruments are played. In addition, we have seen a standardization to follow twelve-tone equal temperament.

The professionalization of instrument makers and uniformation of instrument types have led to the standardization of today's Western traditional instruments. It may be easy to forget that these instruments have been been through numerous development iterations. There is—and has always been—a lot of experimentation with musical instrument designs. This experimentation is culturally dependent. Magnusson (2021, 177) cautions against thinking about universality:

> ... music is *not* a universal language: music changes over time, it travels and adapts to cultural contexts, and it is deeply rooted in instruments and other musical technologies. All over the world, there are house-holds whose family members do not understand each other's music (for example, a dad not understanding his teenage daughter's music and vice versa) and very often this rift is caused by the distinct technology used in the production and consumption of the music.

Instruments can be seen as a cultural object that can be explored by anyone. New musical ideas are created in the meeting point between an instrument and a performer. Magnusson (2018) calls this the "ergodynamics" of the instrument, to explain an instrument's expression potential conveyed through its possibilities and limitations. Each performer will bring their own musical and cultural background into the exploration of a new instrument. The result will again influence others. An ethno-organological approach is based on understanding the culture an instrument comes from, how it moves geographically, and how it is used in various other cultures. This involves both historical and techno-cultural investigations.

In a similar line of thought, Morreale (2021) calls for reflection on the neocolonialist tendency of contemporary music technologies. While his discussion is focused on the use of artifical intelligence in music creation, the same can be said about the way digital music technologies spread around the world. For example, most electro-acoustic instruments and digital audio workstations are currently based on twelve-tone equal temperament and grid-based metric structures. This means that they are incompatible with music cultures that rely on microtonal or microrhythmic structures. It is critical that we are aware of the cultural implications of technological developments. Making more flexible music technologies is one way of allowing for more musical diversity.

II

Embodiment

4

Music-Related Body Motion

"What do you mean by 'gesture'?" This is still my standard question whenever I hear the word used in a musical context. The term "gesture" is frequently used but seldomly defined. The same is the case for several other key terms used to describe the role of the body in musicking. This becomes particularly tricky when working between disciplines. Researchers from, for example, musicology, sociology, performance studies, gesture studies, human-computer interaction, and human movement science use many of the same terms but often with a different meaning. When I first started reading up on the literature in the various fields, I became increasingly confused. Every publication seemed to use different terminology to describe the same phenomena. For example, does a "piano gesture" describe physical motion, the experience of an action, a sonic phrase, or something else? And, when talking about "movement," does this relate to what can be measured with a motion capture system? In the following, I will define and describe some core terminology.

Motion

I prefer the term "music-related body motion" when generally describing human bodily activity in a musical context. Such motion is not necessarily "musical." For example, the motion of a pianist scratching her head while playing is not musical, but it would still be categorized as music related. I include "body" in the term to avoid ambiguity with other types of music-related motion, such as moving sounds in a spatial audio setup. Finally, "motion" is used in a traditional physics-based sense: the act of moving. In English, two words describe this phenomenon: "motion" and "movement." They are often used interchangeably, although they have slightly different meanings. The *Oxford Dictionary* suggests this differentiation:

Motion: the action or process of moving or being moved

Movement: an act of moving

In my mother tongue, Norwegian, we only have one word to describe the act of moving ("bevegelse"). The same is the case in many other Germanic languages. The English words come from Latin, where there are also two different words ("motus" and "motio"). Confusingly, these two words are usually defined as being the same (Jacob 2019). In Italian, on the other hand, there are two distinct meanings of the words "moto" and "movimento."

Since I do not have an innate feeling for the difference between motion and movement, I often ask native English speakers to explain how they perceive the difference between them. One explanation I got was to look at an old-school clock. The hands of the clock display movement, and this movement is the result of motion. Then one could reason that movement is more related to how the phenomenon is experienced.

Jacob (2019) suggests that different "perspective views" leave flexibility to the way terms are used. This makes it possible to differentiate between linguistically similar yet perceptually different phrases, such as the "movement of the heart" and the "motion of the heart." The former would be a nonspecific and metaphorical description, while the latter relates to the physical displacement of a particular heart. One could say that the "motion of a pianist's hand" can be measured with a motion capture system, while the "movement of a pianist's hand" refers to the experience of how the hand is moving through the air.

Motion thus appears to be the most scientific term and is also the preferred term in physics and its subfield, mechanics. Much physics terminology is used in biomechanics, the field that studies the motion of living organisms. However, in human movement science, there seems to be a mix of the terms "motion" and "movement." This may be because human movement science is a field on the intersection between biomechanics, psychology, medicine, and neuroscience. Then there is also a need to describe the body using both physical and psychological terms.

Within biomechanics, two core areas of study are kinematics and dynamics. The kinematics describes how fast and in which direction an object is moving. We know from school physics that motion can be defined as the continuous displacement of an object in space over time. This assumes that one knows an object's "position" in space, usually represented in a three-dimensional Cartesian space. It is possible to quantify motion by measuring the total "distance traveled" or "displacement" of an object. These are not necessarily identical. For example, if a person walks in a circle, the distance traveled will increase over time. However, the total displacement will be zero each time the person passes the starting point.

How fast an object moves is usually measured as its "velocity," which is a vector quantity equivalent to an object's "speed" and direction. One can say that the speed is related to the distance traveled while the velocity is associated with the displacement. An object's "acceleration" is calculated as the derivative of its velocity. It tells about how much the velocity changes over time. The acceleration helps identify when an object starts and stops and is a measure that is frequently used in (musical) human–computer interaction. It is also possible to look at the "jerk" (the third derivative of position) and the "snap" (the fourth derivative of position). These may seem like abstract concepts, but studies have shown that we can experience both jerk and snap when the acceleration changes (Eager et al. 2016).

Motion Capture

There are many ways to measure motion, but these can be broken down into two main types of systems: camera based and sensor based (Jensenius 2018). Some systems measure an object's position in space, while others measure velocity or acceleration. It is easy to find an object's velocity or acceleration by calculating the first or second derivative of position. It is more challenging to go the other way, that is, calculate position from acceleration.

Since camera-based systems are better at measuring an object's position in space, these are often considered state of the art among motion capture systems. The most reliable systems use multiple infrared cameras positioned around the capture space (figure 4.1). The cameras emit infrared light, which is reflected on markers placed on the body of the person or object to be captured. The reflected light is then captured by the cameras and sent to a computer that calculates the exact position through a "triangulation" process. The result is three-dimensional tracking of each of the markers in space. The benefit of such systems is that they can capture at high speeds (more than 100 Hz) and high spatial resolution (in the range of millimeters). The captured points can be visualized directly or used as the basis for further analysis.

While infrared, marker-based systems are considered state of the art, there are also other camera-based motion tracking solutions. New

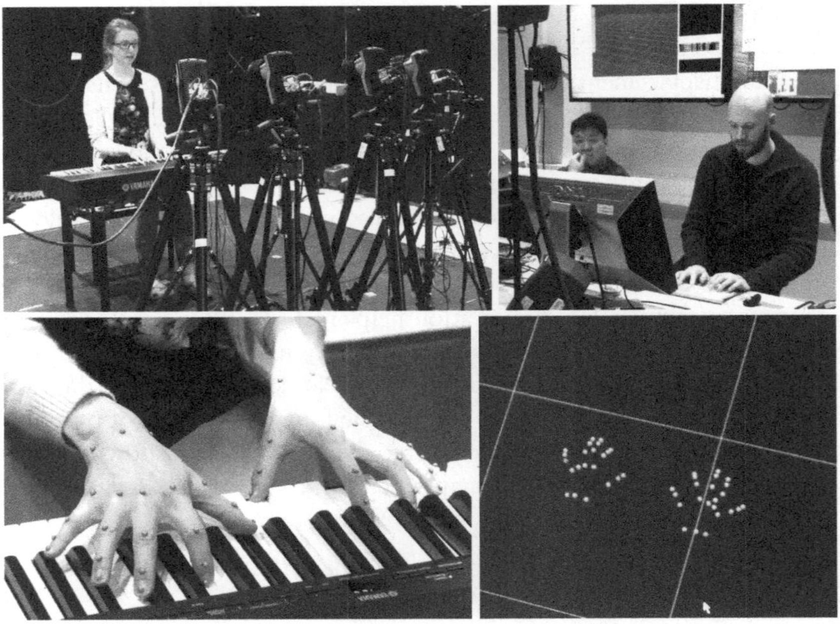

Figure 4.1
Setup for optical infrared motion capture of pianist Christina Kobb in the fourMs Lab at the University of Oslo. There are multiple cameras placed in front to allow for capturing the small markers placed on the hands (bottom left). The markers can be visualized in a three-dimensional grid in software (bottom right).

computer vision techniques makes it possible to extract features from regular video recordings (Moeslund and Granum 2001; Moeslund et al. 2006; Rautaray and Agrawal 2015). Also, new camera types help push the field forward. This includes stereo cameras that mimic the function of the human eyes and depth cameras that can measure the distance to objects. New artificial intelligence–based models also allow for capturing full-body motion with a moving camera (Voulodimos et al. 2018).

There are several drawbacks of using camera-based motion capture systems. The price tag is one, although such systems have become more affordable in recent years. More problematic are the constraints enforced by the need to have a camera focused on a particular recording space and that the cameras need to "see" what is going on. This often leads to occlusion, meaning that certain areas or markers are invisible to the cameras (for example, a dancer moving close to the floor or a cellist whose instrument covers a large part of the performer's body). Then it is impossible to capture the motion of the whole body accurately.

Many of the problems with camera-based motion capture systems are nonexistent in sensor-based systems. There are numerous such systems, based on various types of sensors: acoustic, mechanical, magnetic, inertial, and electrical (Bishop et al. 2001). Of these, the inertial ones are the most popular for measuring motion. These days, inertial measurement units (IMUs) are almost ubiquitous. These sensor units contain accelerometers that measure the gravitational pull. This signal can measure the rate of change of motion, although it is not identical to measuring acceleration as the second derivative of the position. This also makes it difficult to calculate the position from accelerometer data without considerable drift (Skogstad et al. 2011). IMUs typically also contain gyroscopes and a magnetometer. A gyroscope measures the rotation of the sensor, and the magnetometer acts as a compass. Together, these three IMU sensors—accelerometer, gyroscope, and magnetometer—allow for capturing both three-dimensional motion and rotation in a small sensor unit.

One compelling feature of inertial sensors is that they are not affected by external factors such as ferromagnetic objects or light. They are also small and cheap, use little power, and can be sampled at high frequencies. Even though they are ubiquitous in consumer electronics, they are not embedded in many commercial musical instruments (yet). However, as reviewed by Medeiros and Wanderley (2014), many researchers explore their potential in musical applications.

Force

In the previous section, we only considered the dynamics of moving objects. However, for objects to begin (or keep) moving, there needs to be some force acting upon them. Newton's second law tells us that the force (F) is related to the product of a body's mass (m) and acceleration (a): $F = ma$. This means that a massive object requires more energy usage to move than a lighter one. Newton's third law tells that two objects exert equal force on each other for objects at rest. This is relevant from an interaction perspective since the pressure (p) that you use to push on something is related to the amount of force (F) used and the area (A) that it is used on: $p = F/A$.

The force applied to an object can be measured with a force-sensing resistor (FSR). Such sensors are becoming increasingly popular in new musical interfaces since they allow for continuous control of sound features. We will look more at some examples of such usage in later chapters. Force can also be estimated from measuring muscle activity, which has become popular in experimental music devices (Tanaka 1993; Pérez and Knapp 2008). Muscle activity can be detected with electromyograms (EMGs), and it is particularly effective to use EMG sensors on a musician's arms to pick up information about hand and finger motion (Nymoen et al. 2015; Erdem et al. 2020). We will look more at other examples of muscle-based performance in chapter 12.

Another term from school physics is friction, which in its purest form can be defined as $f = \mu N$, where f is the friction force, μ is the friction coefficient, and N is the normal force. Friction is particularly interesting from a musical perspective since it is often the starting point for sound production. For example, the sound of a bowed string stems from the friction between objects and the energy put into the object's motion (Guettler 2002; Serafin 2004; Schoonderwaldt 2009).

Motion and force are also both related to haptics, meaning the sense of touching or being touched. Haptic experiences may be said to consist of both the sense of material properties (tactics) and the motion properties (kinesthetics) of the objects (Cook 1999). While haptic experiences are a natural part of acoustic instruments, few electro-acoustic instruments allow for such a sensing modality. Fortunately, several researchers explore haptics in new instrument designs (Papetti and Saitis 2018).

Perceivers do not generally get a haptic experience, but there have been some experiments with creating haptic-based installations. One example is a series of multimedia installations by Stenslie (2011), in which he explored haptic bodysuits. Turchet et al. (2019) tested giving vibrotactile gilets to perceivers in a music performance context. They found that some were positive to such an extended musical experience, while others were not. Clearly, more research and experimentation are needed to figure out if and how haptics can be used to give perceivers a richer musical experience.

Degrees of Freedom

In engineering, one often talks about "degrees of freedom" (DoF) as the number of independent movement variables in a mechanical system. In a typical three-dimensional scenario, a moving object has six degrees of freedom: three translational (X, Y, Z) and three rotational (yaw, pitch, roll) (Huston 2008). However, not all degrees of freedom may be available in all contexts. For example, a hinge has one degree of freedom since it can only rotate around one axis. A slider has two degrees of freedom, and a ball-and-socket machine joint has three.

In instrument acoustics, the concept of degrees of freedom has been used to describe the instrument's sound-producing elements and how they are connected. Chaigne and Kergomard (2016) argue that a guitar string can be modeled with two degrees of freedom. This means that the string has one degree of freedom and the resonator (the guitar's body) also has one. This simplifies a complex interaction but helps explain the core parts of the instrument's interaction.

In a biomechanical system, like the human body, the joints usually work together. To understand the system's motion potential, it is necessary to add up the degrees of freedom from the involved joints. The human body has seventeen large parts, and these give us a total of fifty-four degrees of freedom when moving around (Huston 2008). With all the smaller joints of the body, there are approximately 230 joints in total. Since there are so many variables to consider, it is difficult to calculate the total number of degrees of freedom of the body. In any case, it may be more interesting to look at the degrees of freedom of the interaction in question.

The degrees of freedom of a (bio)mechanical system is related—but not similar—to the system's number of "control dimensions." By control dimensions, I mean the degrees of freedom of a physical instrument that corresponds to the degrees of freedom of the performer's body parts. For example, a one-dimensional finger joint cannot exploit the three-dimensional possibilities of a ball-and-socket-based joystick. Therefore, the control dimensions of an instrument will be lower than the separate degrees of freedom of the instrument and the performer. Take the case of the cello, an instrument that allows for a high level of "expressivity." The cello is an instrument that is considered difficult to master. However, Rowe (2001) argues that a cellist only has four degrees of freedom in performance: bow force, bow velocity, bow distance from the bridge, and finger position on the fingerboard. I would argue that string selection could be a fifth control dimension, but we are still talking about a low number of control dimensions. So the complexity of mastering to play the cello is not due to its degrees of freedom. Similarly, developing instruments with higher structural complexity does not automatically lead to "better" interaction.

Calculating the control dimensions of an instrument is not a straightforward procedure. For example, one could argue that an individual piano key only has one degree of freedom: key down and key up. The keys are touch sensitive, so the velocity with which one hits the key directly influences the sound level. There are many keys to choose from, but selecting an individual key could be considered one more control dimension.

The pianist, therefore, has two control dimensions: which key to play and the velocity and force used in the attack. These dimensions are called "pitch" and "velocity" in the MIDI protocol (we will get back to a discussion of MIDI in chapter 6). In addition to the keys, the piano also has two or three pedals that can alter the sound level and timbral qualities.

The organ has a similar keyboard interface to the piano, but its control dimensions differ. An organ player also makes a key selection but does not have velocity control on the keys. The sound level is controlled separately from the keys by adjusting the airflow to the organ pipes. On the other hand, in the organ, it is possible to change the timbre of the sound by adding or removing pipes. So even though the piano and organ share the same pitch control interface (the keyboard), they differ in the other control dimensions (sound level and timbre).

As we shall see in chapter 10, new keyboard-based interfaces allow for multidimensional control on each key. This radically changes how a keyboard-based instrument can be played. An interesting question then arises: Is the total number of control dimensions that much different in practice? The limited number of control dimensions on each piano key is compensated by playing many keys simultaneously. It is no problem to play chord progressions with five to eight keys on the piano. My experience is that I tend to focus on only playing on one to three keys simultaneously on multidimensional keyboards. I do not have the cognitive capacity to control more than a few multidimensional keys continuously.

Action

The control dimensions of an instrument is an example of how we can apply a techno-cognitive approach to the analysis of an interaction between a human and an instrument. This is based on investigating what can be measured objectively (the number of instrument or body parts) and how these can be used in performance. Action is another techno-cognitive term. It is related to motion, but it is not the same. Motion describes the physical displacement of an object in time and space. This displacement can be measured objectively with a motion capture system. In my thinking, motion is a continuous phenomenon. Everything is always in motion, from the atomic to the galaxic level. In nature, motion does not have a beginning and end. Therefore, it does not make sense to talk about "a motion." As soon as we segment the continuous motion signal into a temporal segment, we interpret it as a time-limited action. Action is a subjective entity.

Let us carry out a small thought experiment to clarify the difference between motion and action. Imagine a drummer performing an individual stroke on a snare drum. You will easily be able to identify this as a single action: the performer lifts the stick, hits the snare drum, and returns to rest. If performed in a continuous flow, it would be perceived as one coherent action. If you were to measure this action with a motion capture system, however, you would get a continuous signal describing the position of the drummer's hand. When does the action begin and end? This is a matter of interpretation. The motion capture recording will display a physical motion representation from the beginning to the end of the capture session. There is nothing in the motion capture data that will inform about the action. You may look at the graph and locate where you believe that the action began and ended. However, this is your subjective interpretation of the observed motion. The motion capture data can reveal the exact height of the stick, but it will not tell when the action started and stopped.

I think about an action as a "spatiotemporal" phenomenon. Space and time are merged in my mental representation of the action in question. To have a way of representing this on paper, I started experimenting with video visualization techniques (Jensenius 2013c). One example of a video visualization can be seen in the "motion history images" of a series of individual drum strokes in figure 4.2. I made this visualization from a continuous video recording of individual drum strokes. The segmentation into separate images is based on a threshold applied to the "quantity of motion" calculated from the video sequence. The assumption was that an action would start after the percussionist had stood still for a second or two between strokes. The segmented video clips were "averaged" into one motion history image. Each of these images then represents one action, at least according to my interpretation of the sequence. Time is indirectly represented through the spatial and temporal distribution of the motion. It is impossible to see precisely when something happened, as when plotting motion capture data along a time axis. But you get a compact visual representation of the action as a coherent spatiotemporal unit.

As the above example showed, an action may be understood as a holistic unit—or a "chunk"—of motion or force (Buxton 1986). Such an action chunk is subjective, meaning that people may perceive actions differently. In cognition in general, and music cognition in particular, we may talk about three temporal levels, each related to the three main memory functions: sensory, short-term, and long-term memory (Snyder 2000). Based

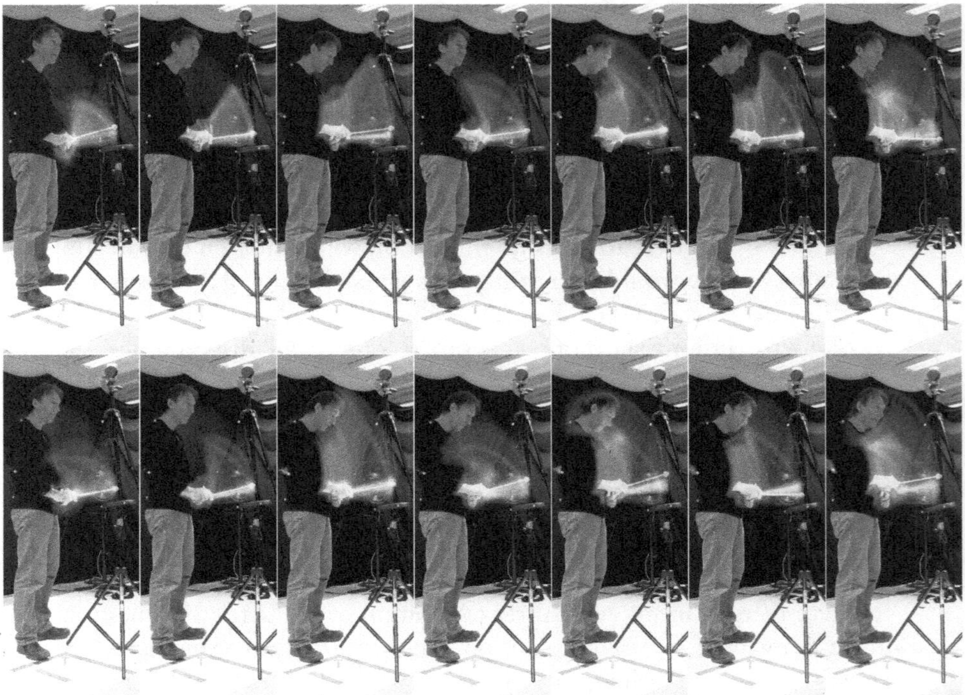

Figure 4.2
Motion history images created from a video recording of a percussionist performing individual drum strokes on a pad. Each image represents about fifteen seconds of video material and can be seen as a way of representing the spatiotemporal characteristics of the individual actions.

on such a tri-part division, Godøy (2008) has suggested three levels of chunking:

- **Subchunk level:** The perception of continuous sound and motion features. This is based on sensory memory and covers events up to 0.5 seconds.
- **Chunk level:** The perception of fragments of sound and motion that are perceived holistically, such as sound objects and goal-directed actions. This is based on short-term memory and is typically between 0.5 and 5 seconds.
- **Suprachunk level:** The perception of several chunks concatenated into larger structures. This is based on long-term memory and covers events longer than 5 seconds.

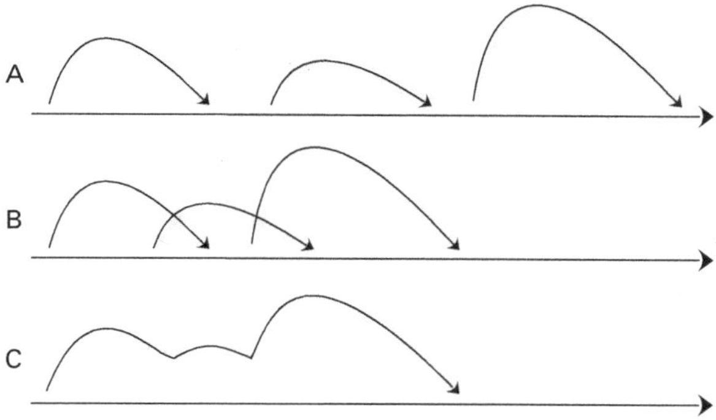

Figure 4.3
Sketch of motion over time of individual goal-directed-actions (A) and how they will overlap if performed so close together that their prefixes and suffixes overlap (B). This would result in the coarticulation of the action trajectories (C) (based on Godøy 2014).

Humans handle these levels effortlessly and in parallel. For example, we may observe the instantaneous unfolding of sound and motion while at the same time keeping an internal memory of the trajectories of a sequence and an overall image of extended patterns. Analyzing the same with a computer-based system reveals the complexity of this multilayered parsing task. Particularly challenging is that action chunks are context dependent, meaning we perceive them relative to what happened before or after. They can also be "nested"; several actions that follow each other may be perceived as one large action. For example, playing a scale run on a piano can be considered a series of separate actions if we focus on each finger's motion. If we instead look at the motion of the hand or the upper body, a scale run can be perceived as one coherent action chunk. Then we may talk about the effect of "coarticulation" (Hardcastle and Hewlett 1999), the merging of individual units into larger supraunits (figure 4.3).

From an interaction perspective, we may think of actions as "foreground" activities in a continuous motion stream. Then it also makes sense to talk about a "background," motion outside our attention. This is sometimes called "fidgeting" and can be defined as actions that are peripheral or nonessential to the main task (Mehrabian and Friedman 1986). We may even think of fidgeting as "motion noise," which is typically filtered out when one deals with action recognition. That said, looking into noise can reveal relevant features. I have for some years been interested in human "micromotion," the tiniest motion that humans can produce and perceive. This I have done through motion capture studies of people standing still

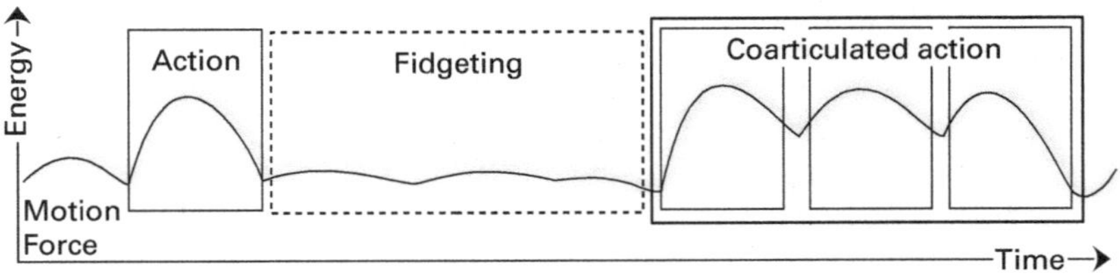

Figure 4.4
A sketch of the relationships between motion/force, action, and fidgeting. Motion is the continuous displacement in time and space. Actions are subjective motion chunks, while fidgeting refers to the "motion noise" between actions. Actions may coarticulate into larger action units.

while listening to silence and music (Jensenius 2017a; González Sánchez et al. 2018, 2019; Zelechowska et al. 2020). In these studies, the participants were explicitly told not to do anything, just to stand still. The aim was to look at the motion of their standstill and see if we could find systematic behavioral patterns. We are here talking about motion at a scale up to around 10 mm/s. We have found that there are, indeed, systematic differences in people's motion patterns. The nature of motion and its relationship to musical features is still under active research. My scientific investigation into human micromotion is tightly connected to artistic explorations into sonic microinteraction, which we will look at in chapter 12.

Figure 4.4 is an attempt to summarize how I think about the differences between the objective terms "motion" and "force" and the subjective concepts "action" and "fidgeting." Note that an action can be based on either motion or force. It is possible to see muscle tension without measuring any motion. For example, try pressing your fingers against a table. You can feel the exertion of force, but there will be no motion. This is an example of a motionless action. Similarly, there may be motion without any exerted force. However, the most important is distinguishing between what is measurable (motion and force) and what can only be inferred from the measurements (action and fidgeting).

Gesture

The last term to define at this stage—and also the most complex one—is "gesture." This is a term that has received growing interest in recent decades among music researchers (Gritten and King 2006, 2011). Around twenty years ago, Cadoz and Wanderley (2000) wrote a summary of different types of gesture definitions in music and music interaction. I contributed with an updated review ten years later (Jensenius et al. 2010) and a study of how the term "gesture" was used in the music technology community (Jensenius 2014). The main finding from these review studies was that the term gesture is used broadly. Some use it more or less synonymous with what I call motion, others with action, and others again in a more linguistic sense.

I prefer to think about gesture the way it is used in linguistics and behavioral studies, what Kendon (2004, 7) refers to as "visible action as utterance." In my thinking, a gesture is related to conveying "meaning" through action. Meaning should here be understood broadly and can—as we will get back to later—also be expressive. The main point is that a gesture is not only physical motion or any goal-directed action; it is also part of human communication. McNeill (1992, 12–19) proposed a taxonomy of different gestural functions:

- Iconics represent a particular feature of an object and can be described in terms of the gesture's shape and spatial extent. Iconic gestures are often used to illustrate an action, such as imitating a knocking movement with a hand while saying "knocking on the door."
- Metaphorics are similar to iconics but represent an abstract feature of an object. An example of a metaphoric gesture may be saying "something happened" while holding up the hands to refer to "something."
- Beats occur together with spoken words to highlight discontinuities and to stress specific words. Beats are typically carried out as in/out or up/down actions and may be seen as emphasizing the most important words in a narrative.
- Deictics indicates a point in space (for example, pointing in a specific direction while saying "over there").
- Emblems are stereotypical patterns with an agreed meaning, such as waving "goodbye" or the "thumbs up" sign.

McNeill's gesture theory is built on the idea that gestures coexist with speech. This is not to say that they have to co-occur, but instead that gestures and speech are "coexpressive." Here, McNeill (2005, 15) argues that "language is inseparable from imagery" and that mental imagery is embodied in the gestures that co-occur with speech. To explain the relationships between gesture and speech, McNeill (1992, 37) outlines what he calls the "Kendon continuum," based on the typology of gestures suggested by Kendon (1980): gesticulation, emblems, pantomime and sign language. As shown in figure 4.5, this continuum covers two extremes:

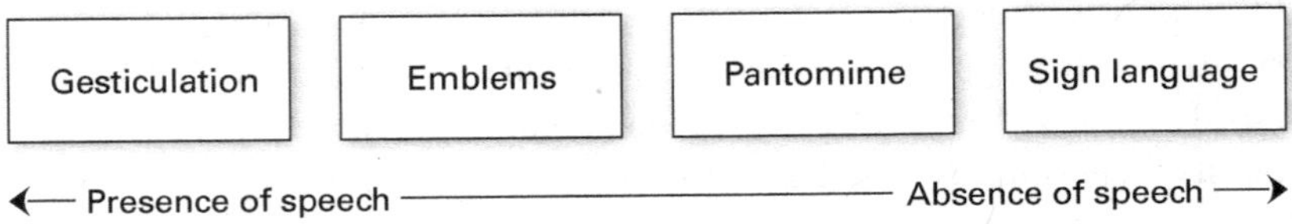

Figure 4.5
A sketch of the "Kendon continuum" shows how different types of gestures relate to speech (McNeill 2005, 7).

gesticulation denotes gestures that always co-occur with speech and sign language the linguistically self-contained ones.

Musical Gesture

The literature on relationships between gesture and speech is relevant when moving into the realm of music. In the music literature, gesture is used in many different ways. For example, Bielawski (1979) proposed a model in which he refers to an instrument as a "transformer" of gestures. While there are also others that use gesture more or less synonymous with motion or action, most think about higher abstraction levels. Métois (1997, 16) proposes this definition:

> [B]oth [physical and auditory gestures] present the ability to communicate musical intentions at a higher level than an audio wave form. The similarity of their level of abstraction motivated the author to label them both as Musical Gestures.

Hatten (2004, 95) argues that a musical gesture is "significant energetic shaping through time." He primarily refers to the experience of musical gesture "within" music, either through the score or the musical sound. Then I find it easier to relate to the definition proposed by Gritten and King (2006, xx):

> [A] gesture is a movement or change in state that becomes marked as significant by an agent. This is to say that for movement or sound to be(come) gesture, it must be taken intentionally by an interpreter, who may or may not be involved in the actual sound production of a performance, in such a manner as to donate it with the trappings of human significance.

This definition suggests a flow of communication between the performer and the perceiver. The performer's motion "becomes" a gesture if the perceiver understands it. An interesting question then arises regarding consciousness: Does an action have to be carried out consciously to be seen as a gesture? From a communication perspective, Kendon (2004,

15) argues that gestures are intentional and therefore have to be carried out consciously. On the other hand, Gritten and King (2006, 162) suggest that a musical gesture can be perceived as significant even if performed unconsciously. This is related to how Murray-Rust and Smaill (2011) talk about "musical acts." In their model, a musical act needs to (a) be embodied, (b) have an intention, and (c) be intelligible. If one person acts and is not understood by another, then it fails to impact. I would argue that we should allow for ambiguity. The same action can be perceived as either intentional or unintentional by different people. In any case, we have to think of a gesture as a subjective entity.

In my thinking, the gesture concept boils down to what in traditional Saussarian semiotics would be seen as a separation between "sign" and "signifier." We can separate between the meaning of a gesture—broadly speaking—and the motion involved in communicating that meaning. For example, think about the gesture of waving "goodbye" to someone. This gesture can be performed with or without speech, but it would nevertheless contain the meaning of saying "goodbye." You can wave with the left hand or right hand, or both. The physical motion would be different, but the meaning (hence the gesture), would be the same.

One prominent example of a musical gesture can be seen in cartoon films. Think of when a character falls off a cliff accompanied by a long glissando illustrating the downward motion of the character, possibly ending with a massive "bang" when the character hits the ground. Here, meaning is attached to the sound, which is also directly related to visible (or imaginary) action. The sound may evoke motion imagery, even if nothing is seen on the screen.

There are also examples of how meaningful motion may lead to sonic imagery. Think of a pianist playing the final notes of a classical piano concerto. The performance ends with a majestic final chord, which the pianist holds while listening to the decaying sound. The pianist may continue to hold the chord long after the sound has become inaudible, say, for three more seconds, before finally lifting the hands high up in the air and back to rest position before taking the applause. Such a motion-based gesture creates suspense and prolongs the experience of the piano sound.

Figure 4.6 sums up how a musical gesture can be related to the continuous motion and sound on the physical side and the perceptual actions and sound objects on the perceptual side. In some cases, a musical gesture may be driven primarily by sound, like in the cartoon case mentioned above. Other times, a gesture may be primarily connected to motion, like in the

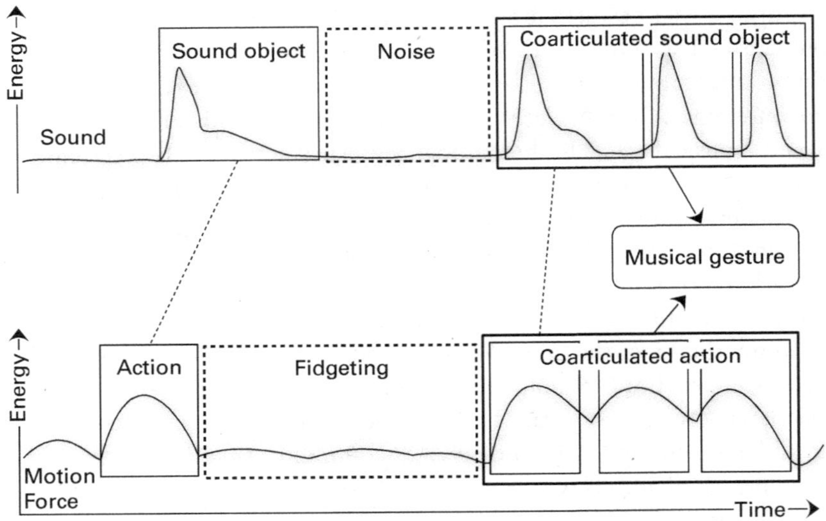

Figure 4.6
A visual summary of relationships between the continuous flow of motion and sound, the chunking of motion and sound into actions and sound objects, and how the two may be experienced as a musical gesture.

piano example. However, in most cases, musical gestures are based on the combination of sound and motion.

Position, Posture, and Pose

The concepts of motion, action, and gesture are all related to moving bodies. They are also useful to describe bodies that are not in motion. Here I suggest to differentiate between position, posture, and pose. Position is the most technical of these, which I will use to describe the physical location of a body, a body part, or an object in space. The position is something that can be measured with a motion capture system. Then we can look at the motion capture data and find that a person's foot was located on the floor 1,157 x 432 x 920 millimeters from the center point in the XYZ coordinate system used in the space. Looking at different points in time, it is then possible to find the various positions that a person has been standing in over the recording.

I will use posture as a subjective term to describe how a person holds their body. A person's "base" posture defines how they stand and can be easily recognized. Posture is also something that one can develop over time through exercise. Think, for example, about how you can learn to make a "power posture." It may be possible to calculate the spatial features of a posture based on position data, but this would require a subjective placement/selection of the points from which the individual positions were captured. So posture is related to the position in a similar way that action is related to motion.

Finally, I will use the term "pose" to refer to postures with a meaning-bearing component. They are often intentional and are typically something that is used in performance. As such, one may say that a pose is based on communication without moving. This is similar to a gesture, with the main difference being that gestures are dynamic while poses are static. Poses also typically refer to how the whole body is organized, although one may argue that setting up a pose with only the hands or head is possible. Lifting your nose, for example, can be seen as a pose to behave "snobbish." There may also be cases during which a pose and a gesture overlap. Many people would probably consider a "thumbs up" sign to be a gesture. I would agree with that if it is dynamic, that is, if you are lifting your hand in a particular way to show the thumb. However, if you only consider the "thumbs up" as a static sign, I would instead call it a pose.

My main point here is that both gestures and poses need to be interpreted to be understood. So-called gesture or pose recognition systems will need to start by capturing position or motion. Then one needs to segment these continuous data streams into postures and actions before finally trying to infer their meaning-bearing components. In the case of "musical gesture systems," one also needs to consider the sound. First, the continuous sound needs to be recorded before it can be segmented into "perceivable" sound objects. Finally, these sound objects can be combined with information about actions to understand the meaning of a musical gesture. This is a nontrivial task, and many researchers work on this topic from different angles. We will look more at some of my explorations in a discussion of air instruments in chapter 12.

5

Functional Aspects

"And then you do that . . . 'thing' . . . in the air . . ." I still remember one of the first rehearsals of *Laserdance*, a dance piece with interactive sound that I was part of setting up. In the piece, the dancers could control the sound by moving in the air. None of us had a vocabulary to describe such sound-producing actions and resorted to talking about the "thing in the air." I realized that we had to work on our terminology. In the following years, I spent quite some time reading up on the literature and trying to come up with a coherent taxonomy for describing the functional aspects of various types of music-related body motion (Jensenius et al. 2010). The taxonomy builds on the work by Gibet (1987), combined with the proposals by Cadoz (1988), Delalande (1988), and Wanderley and Depalle (2004). When I first developed my taxonomy, I used the term "gesture" extensively. After working more on the descriptive terminology presented in the previous chapter, I have decided to use gesture more sparingly. There are times when it makes sense to talk about musical gestures, that is, actions with some intended/perceived meaning-bearing component. However, in many cases, it may be better to speak of motion or action.

Four Main Types

We can separate between four main functional categories of music-related body motion:

- "Sound-producing" actions are the ones that produce sound. They can be subdivided into "selection," "excitation," and "modification" actions.
- "Sound-facilitating" motion or actions are indirectly involved in the sound production. They can be subdivided into "support," "phrasing," and "entrainment" actions.

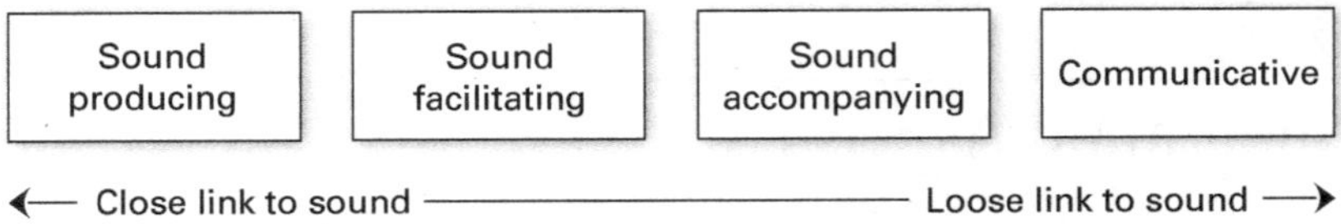

Figure 5.1
A sketch of the relationship between the four different music-related body motion types and their connection to sound.

- "Sound-accompanying" motion or actions are not involved in sound production but follow one or more sound qualities. This includes "sound tracing," "sound mimicking," and "air performance" actions.
- "Communicative" gestures are primarily extramusical. These may be "endogenous," "performer–performer," and "performer–perceiver" types of communication, ranging from being communicative in a linguistic sense ("emblems"), to being examples of more abstract forms of communication ("expressive").

These categories are not mutually exclusive. In most cases, they overlap in various ways. It may help to think of them as placed along an axis, as sketched in figure 5.1. This axis is inspired by the Kendon continuum (McNeill 1992) and shows that sound-producing actions are, by necessity, closely linked to the sound. The other types are more loosely (if at all) connected to the sound. In the following, we will look more at each of the four main types.

Sound-Producing Actions

The first part of performing a sound-producing action on an instrument usually involves making a "selection" of which key to press, which string to hit, and so on. This happens continuously during a performance on many instruments. There are also examples of selection actions performed before performing, such as changing the stops on an organ. Instruments constructed with many mechanical parts often afford more selection actions. One could argue that much of the preparation done in electroacoustic instruments—such as selecting which presets to use or loading samples—could be seen as selection actions.

The actual sound production is happening with an "excitation" of the sound-producing element. In its purest form, an excitation action may be either "direct" or "indirect." A direct excitation action is when the performer touches the sound-producing element with the body, such as

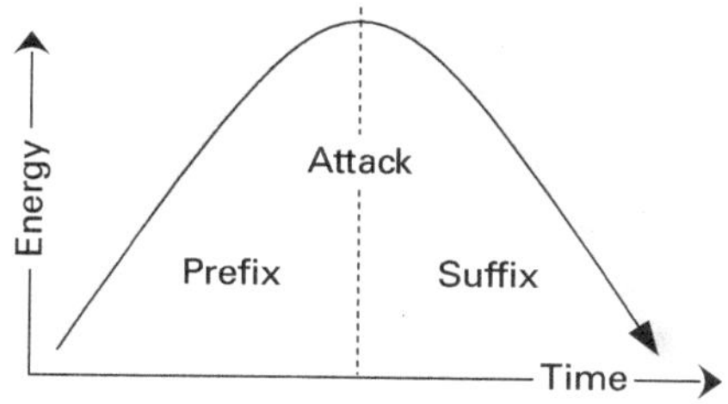

Figure 5.2
The excitation part of a sound-producing action may be seen as having an attack surrounded by a prefix and suffix (Godøy 2008).

blowing on a reed, plucking a string, or hitting a drum with the hand. Indirect excitation actions occur when an object is between the body and the sound-producing element, such as when playing with a bow, pick, or key.

Godøy (2008) suggests that an excitation action can be decomposed into the main "attack" preceded by a "prefix" and followed by a "suffix" (figure 5.2). The attack defines the sound and is largely based on the shape and quality of the prefix. The suffix is a continuation of the attack and typically works best if the performer can obtain fluidity through the whole action from the beginning of the prefix to the end of the suffix. In many cases, the suffix is also the preparation for the next attack. If played rapidly, the suffix of one action may overlap with the prefix of the next action according to coarticulation principles.

The prefix, attack, and suffix are essential parts of a sound-producing action and are important for the perceiver. The prefix may guide the attention and set expectations for the sound to follow. For example, if a percussionist lifts a mallet high above a timpani, one would expect a loud sound. It is also expected that the rebound of the mallet (the suffix) matches the energy level of the prefix. As such, both prefixes and suffixes may help to "adjust" the perception of the sound based on ecological knowledge of the involved objects and actions.

When it comes to describing the quality of the sound-producing action, Godøy (2006) proposes three action–sound types based on the "sound object" classification by Schaeffer (1967):

- **Impulsive:** The excitation is based on a "discontinuous" energy transfer, resulting in a rapid sonic attack with a decaying resonance. This is typical in percussion, keyboard, and plucked instruments.

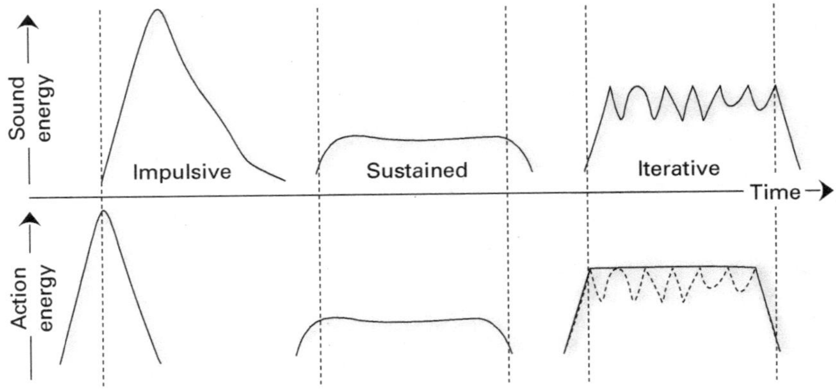

Figure 5.3
Sketch of energy levels for actions and sound objects for the different types of sound-producing actions. The dotted lines indicate the attack phase. Iterative sound objects may result from a series of impulsive actions of the performer or a sustained action by the performer on an instrument that produces a series of impulsive actions.

- **Sustained:** The excitation is based on a "continuous" energy transfer, resulting in a continuously changing sound. This is typical for wind and (bowed) string instruments.
- **Iterative:** The excitation is based on rapid and discontinuous energy transfers, resulting in sounds with successive attacks that are so rapid that they tend to fuse into perceptual chunks. This is typical in some percussion instruments, such as guiro and cabasa. It can also be found when one performs rapid attacks on other instruments, such as using the plectrum rapidly on a guitar.

As shown in figure 5.3, the action–sound types may be identified from the action and sound energy profiles. Sound is produced during the attack phase of the sound-producing action. Thus the prefix and suffix parts of the action are usually soundless, although there may be resonance in the instrument and reverb in the room.

Two action possibilities are sketched for iterative sounds in figure 5.3. Iterative sounds may result from either the instrument's construction or the action with which it is played. Percussion chimes can produce iterative sounds from continuous actions. Here the performer moves the hand with a sustained action over the rods, and it is the interaction of the moving rods that creates the iterative sound. Playing a tremolo on a piano, on the other hand, involves a series of iterative actions by the performer. Here, rapid actions fuse into one superordinate action based on coarticulation principles. In either case, iterative sounds and actions may be seen as having different properties than that of impulsive and sustained action–sound types.

The tripartite model sketched above is, of course, an oversimplification of a complex reality. For example, a violin may be played with many different excitation actions, and, as documented by Applebaum and Lindsay (1986), these may sometimes even merge. Sustained violin tones can be played with "detaché," in which the bow does not leave the string. The different types of "martelé" bowing styles are based on short attacks but without the bow leaving the string, several of which can be categorized as an iterative sound character. In addition comes various types of impulsive sound types, including the finger plucking in "pizzicato" playing. All these different playing styles also lead to different timbral nuances.

The world of electro-acoustic music has opened for even more sonic experimentation, often focusing on creating new timbres and textures. In the following, I will use timbre to denote the sonic quality of a perceived individual sound object, which can be thought of as the sound's "color." I will use texture to refer to the combined sonic quality of multiple sound objects. So we could say that an instrument has a timbre, while an ensemble has a texture. In electro-acoustic music, it is not always possible to separate the two, and timbre and texture may seem as overlapping terms. Many composers and producers deliberately create new sonic timbres/textures that break with our traditional understanding of impulsive, sustained, and iterative sounds. For example, Smalley (1997) argues granular sound is a sound type between sustained and iterative sounds. Working with digital signal processing or synthesis techniques makes it possible to create unnatural sounds, including sounds based on "reversing" traditional impulsive envelopes. As sketched in figure 5.4, it may then make more sense to think of a sound's energy profile as a continuum from more impulsive-like to more sustained-like sounds.

When it comes to the perception of sound objects, recent research by some of my colleagues at the University of Oslo shows that this is more complex than previously acknowledged. Danielsen et al. (2019) showed that the perceptual experience of an impulsive attack—the "perceptual center" (P-center)—is not always easy to determine. Through perceptual experiments, they found that the P-center is usually somewhere before the energy peak. However, this depends on the timbral content and the temporal development of the sound object. Creating a machine listening system to detect such P-centers needs to include an understanding

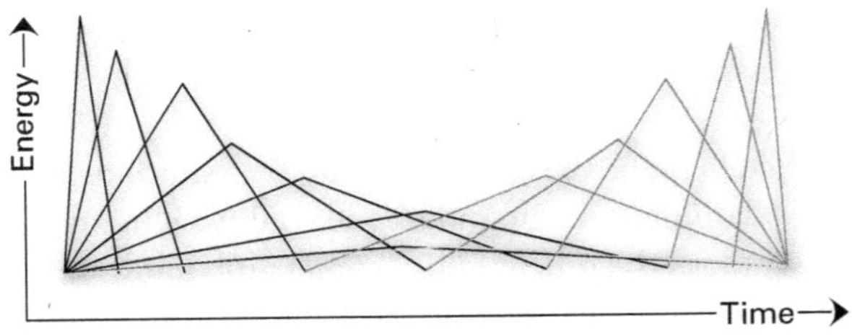

Figure 5.4
A sketch of different types of attack–decay envelopes that can be found in electro-acoustic instruments, ranging from purely impulsive to purely sustained sounds, and including "reversed" sounds.

of the relationships between the differences between the physical sound, its signal-based representation, and the perception of where the attack is located (Lartillot et al. 2021). This is challenging enough for impulsive sound objects and more so for sustained and iterative sounds with long and undulating development.

The third type of sound-producing action is that of "modification" actions. As the name implies, these actions change the quality of the sound in one way or another. (Cadoz 1988) suggests dividing such modification actions into two groups:

1. **Parametric:** Actions that continuously change a parameter, such as a bow pressure in violin playing
2. **Structural:** Actions that modify or change the object's structure, such as opening or closing a key on a wind instrument

Many musical instruments are played with both excitation and modification actions, and, in many instruments, the two types are easily separable. An example is string instruments in which the two hands play entirely different roles: the left hand mainly modifies the sound (choosing the pitch) while the right hand is carrying out the excitation. In wind instruments, the excitation is done in the mouth, and modification actions are carried out with the fingers. There are also cases in which it is difficult to separate excitation and modification actions. Kvifte (1989) argues that there are couplings between these two action types in wind instruments. The mouth can both produce sound and modify its quality. Similarly, string players use the left hand for selecting strings and modifying the pitch through the placement of fingers. However, the left hand can also be used for sound production. For example, string players can produce sound by pressing hard on a string with the left hand.

Sound-Facilitating Actions

The second main category of music-related body motion, "sound facilitating," is not directly involved in sound production. Still, such motion or actions play an essential part in the shaping of the resultant sound. One sub-category here is "support" actions. For example, hitting a piano key involves the hand, arm, and upper body. This multijoint system is necessary to put the finger in motion, such that it hits the right key with a specific velocity at a particular point in time. Support actions may even have audible components. For example, Wanderley et al. (1999) showed that clarinetists' sound-facilitating actions in the upper body had an audible component due to the instrument's changing sound diffusion pattern.

Performers' "phrasing" actions are other types of sound-facilitating actions that directly impact the musical result, although they are not directly involved in the sound production. Wanderley and Orio (2002) showed that the sound-facilitating actions—what they called "ancillary gestures"—of clarinetists are an integral part of the instrumentalists' performance. Such sound-facilitating actions are remarkably stable and reproducible over time, probably because they are internalized as part of learning a piece (Campbell and Wanderley 2005).

The third type of sound-facilitating actions is related to "entrainment," the synchronization of two or more independent rhythmic processes (Clayton et al. 2005, 2020). Entrainment is typically seen in rhythmic music, in which a part of the body—for example, the foot—may synchronize with the music's pulse. Although bodily entrainment varies considerably between performers and performance styles, they may be thought of as necessary for the performance timing (or lack thereof). As such, entrained motion can be a generator of rhythm and timing, in the same way as the rhythm and timing in music can be a generator of motion (Clarke and Davidson 1998).

It is not always easy to differentiate between sound-producing and sound-facilitating actions. Although most musicians would probably not think about the difference between the two in performance, they would acknowledge the difference during rehearsal. I still remember how my first piano teacher asked me to put my hand on her wrist to feel how she moved her hand in a circular motion when playing arpeggios. The sound-facilitating actions are typically more visible than the sound-producing actions and are therefore also crucial for perceivers. Seeing a pianist playing with the elbows out to the sides, for example, is a different experience from holding the arms down along the sides. Furthermore, large phrasing actions in the upper body can be experienced as expressive gestures.

Sound-Accompanying Actions

The third main category of sound-producing actions, sound accompanying, is primarily focused on "following" qualities in the sound. More than 1,500 years ago, Boethius (1989, 8) wrote, "How does it come that when someone voluntarily listens to a song with ears and mind [...] his body responds with motions somehow similar to the song heard?"

This question is equally valid today, and it has received renewed interest in recent music cognition studies. At the University of Oslo, we have contributed with several studies on "sound tracing." This is a type of sound-accompanying motion based on representing perceived sound features in the air with the hand. When hearing a tone with a rising pitch, many people will spontaneously move a hand upward. Coincidentally, many people would also say that the pitch "goes up," an example of a bodily metaphor related to a sonic feature. To understand more about such relationships between people's spontaneous body motion and musical sound, we carried out a study in which we asked participants to "draw" the sound they heard on a graphic tablet (Godøy et al. 2006a). The aim was to see what sound features the participants would respond to and trace with a digital pen. It turned out that people used several different strategies, which could be summarized as either:

- mimicking the sound-producing action(s), such as quickly pressing down for impulsive sounds and "bowing" for sustained sounds,
- drawing one or more sound features, such as the dynamic envelope or the pitch contour, or
- drawing an abstract shape, with some metaphoric content, such as "lifted" or "floating."

Some of these alternatives are related to specific features present in the sound itself. Others focus on imagery of the sound-producing action. Others again were entirely metaphorical. In addition to tracing individual sounds, participants were also exposed to composite sounds consisting of multiple attacks or partly overlapping spectral features. One example of some responses to such a task can be seen in figure 5.5. Here we found, unsurprisingly, that more musically trained participants identified and traced the different parts of the composite sounds better than untrained participants.

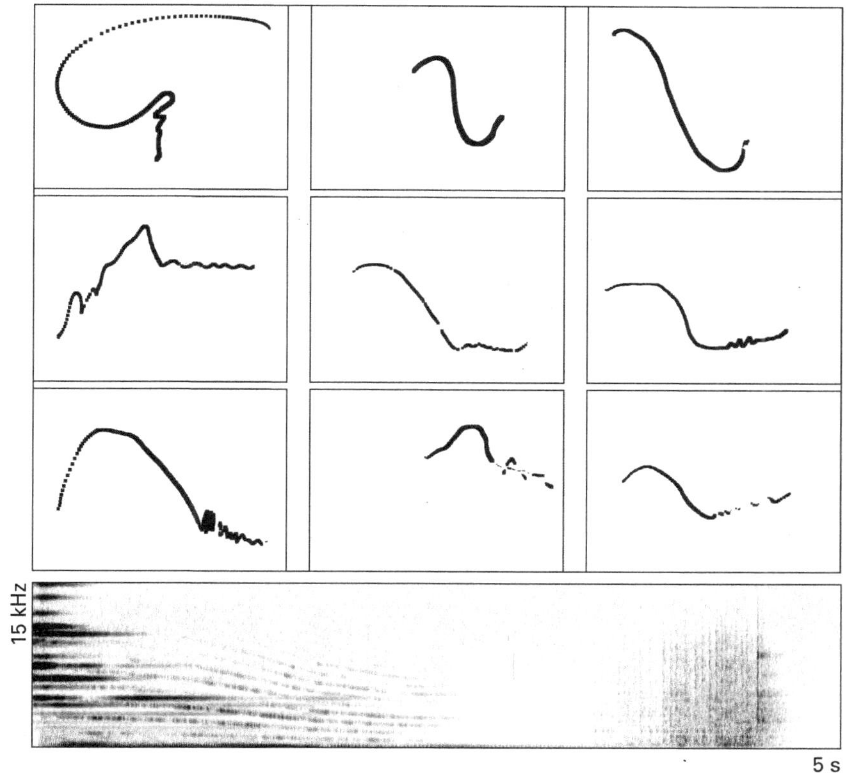

Figure 5.5
Nine tracings to a complex sound, composed of a long, downward-moving slide, followed by a short, rattle sound at the end. Musically trained participants typically traced both of the sonic objects, while less musically trained participants followed only the main envelope of the sound (Godøy et al. 2006a).

Moving the sound-tracing paradigm into three-dimensional space, Nymoen (2013) conducted sound-tracing experiments where the participants would move two hands in the air to short, synthetic sound objects. The motion capture analyses revealed a high correlation between the pitch height of the sound and the hands' vertical position. Kelkar (2019) continued this line of research but with musical material. She carried out a series of sound-tracing studies using speechless, sung melodies from four musical genres: classical vocalize, jazz scat, Sámi folk music, and a North Indian rag. She found that some participants related pitch height to vertical hand position. However, the main finding was that several other motion strategies were also used, summarized in figure 5.6.

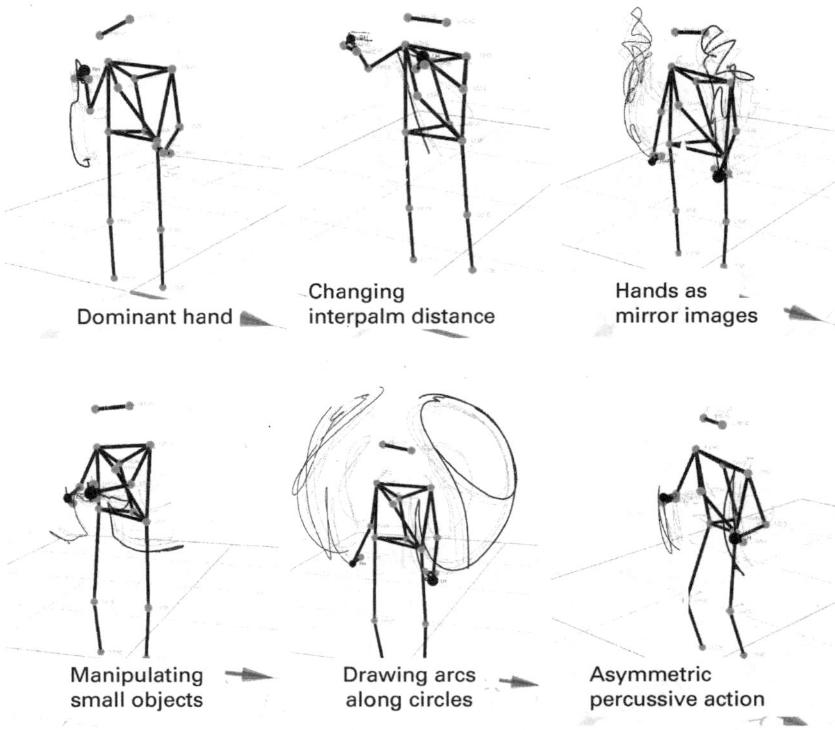

Figure 5.6
Motion history images exemplifying the six dominant sound-tracing strategies in a study by Kelkar and Jensenius (2018).

While most of our sound-tracing studies have focused on individual sound objects or melodies, we have also experimented with sound-accompanying actions to rhythmic music. In a study of people's ability to synchronize a "virtual shaker" to music, participants were asked to move a short wooden stick in the air (Danielsen et al. 2015). As expected, most participants were able to do this effortlessly when the rhythm was clear and easy to follow. However, moving to stimuli with microrhythmic variations was challenging for most participants. Only some musically trained participants managed to achieve a high level of synchronization with these patterns.

Another type of sound-accompanying action we have studied is that of "air performance." While the sound-tracing paradigm primarily investigates peoples' ability to follow sound features, air performance mimics imagined sound-producing actions. In an "air piano" experiment, participants were asked to spontaneously play piano in the air, as if producing the music they heard (Godøy et al. 2006b). The analyses revealed that participants were able to perform the task regardless of musical training. The main difference was that musically trained participants showed a higher level of detail to onsets and register than the novices. The latter primarily followed the music's general contour. Perhaps the most exciting finding was that even people that claimed to be "tone deaf" and "unmusical" managed to complete the task relatively well.

Dancing is another example of sound-accompanying motion, and asking people to dance freely to music can tell you about how they experience the music. In a series of free-dance studies, we looked at how individual dancers follow musical patterns (Jensenius 2007; Haga 2008). Since we asked the dancers to repeat their improvisations to the same musical material three times, we also studied similarities and differences between each participant's performance. These studies found a correlation between pitch height and vertical motion, albeit weaker than in the sound-tracing experiments. More importantly, we found correlations between physical effort and the spectral complexity and loudness of the sound.

Dancing usually happens in a social setting. To understand more about the intersubjectivity of dance experiences, we set out to study how groups of people dance together in a club-like environment (Solberg and Jensenius 2017, 2019). One aim was to understand more about the social dynamics of the groups, which consisted of up to ten people at a time. We were also interested in looking at the effect of the "break routine" typically found in electronic dance music (EDM). This is a moment in which the regular pulse dissolves, followed by a gradual pitch rise accompanied by an increase in the timbral/textural complexity of the mix. The break routine ends when the beat is "dropped" back and the groove continues. We found that the break routine did, indeed, influence the dancers' rhythmic motion patterns. We also found that the reintroduction of the groove after the drop "re-energized" the group.

In parallel to studying large-scale sound-accompanying motion and action, I have run studies on music-related "micromotion" (Jensenius 2017a; González Sánchez et al. 2018, 2019; Zelechowska et al. 2020). Using a "standstill" paradigm, the participants have been asked to stand as still as possible on the floor for six to ten minutes while listening to (musical) sound and silence in randomized order. Of course, it is impossible to stand completely still, and we have been measuring people's micromotion level using an optical motion capture system. By playing different types of music to people when they stand still, we can learn more about the different motion-inducing effects of the music in question. Not surprisingly, we have found that dance music with a regular beat pattern most significantly influences the micromotion. However, how and why this is the case are still open questions.

Communicative Gestures

The fourth main category of music-related body motion is that of "communicative" gestures. I call these "gestures" since they are focused on communicating meaning in one way or another. Communicative gestures may be endogenous, but they are often used to communicate between performers or between performers and perceivers. Such gestures can be communicative in a linguistic sense (emblems) or abstract (expressive). The way a performer communicates with an audience is often critical for how the concert is experienced. Equally important is performer–performer communication, which may involve anything from glancing at each other to support the continuous adjustments of musical features throughout the flow of performance to actions that resemble conductor messages (Timmers et al. 2022). With the advancement of new tracking technologies, it is now possible to empirically study performer–performer communication. For example, eye-tracking studies of a string quartet reveal how the other musicians look at the first violinist throughout a performance (Bishop et al. 2021).

The "conductor" is a musicking actor that has received little attention in this book so far. It may perhaps be more relevant to talk about the "musical leader" to also include people without the institutionalized title "conductor." Such a musical leader can be the first violinist of a string quartet, the founder of a jazz band, the lead singer in a pop/rock band, or the conductor of a full-size orchestra. The focus on gestural communication is what differentiates the musical leader from the other members of an ensemble. Davidson and Correia (2002, 243) describe how Annie Lennox leads her band:

> She is a narrator-interpreter in her use of illustrative and emblematic gestures with the co-performers and audience. She is a co-worker in her use of regulatory movements to coordinate musical entrances and exits.

Leante (2014) argues that khyal singers in North Indian classical music communicate the lyrics through iconics or metaphorics and perform abstract gestures along with the flow of the improvised sections.

However, musical leadership is not always connected to a particular person. McCaleb (2014) describes how the leadership varies within a string quartet, what he refers to as the "fluidity of ensemble roles." While such musical leadership is possible in relatively small ensembles, larger ensembles often rely on a traditional conductor engaging in what Volpe et al. (2016) call "sensorimotor conversations" with the musicians. A conductor can be seen as a communicator of emotional content, besides providing temporal and structural information to the musicians. We will return to the conductor's role in a discussion about the orchestra as "instrument" in chapter 7.

I find it intriguing that new motion-tracking technologies make it possible to perform "in the air." Then it is possible to use communicative gestures for sound production and modification. However, what are meaningful mappings between motion and sound? How does one extract actions from the continuous motion signal? How does one interpret the meaning of gestures from the segmented actions? How does such a shift in function from communication to sound production challenge expectations about how sound is produced and modified? We will return to such questions when discussing air instruments in chapter 12.

6

Representations of Sound Actions

"Why don't you try to map the perceptual features instead?" one of my PhD colleagues asked as I pondered how to control a sound engine with an electromagnetic motion tracking system. Until then, I had mainly thought about mapping as a technical procedure. You acquire some control signals that are processed a little before routing them to a feature in the sound engine. I had not thought about the possibility of creating higher-level or multilayer mappings. It occurred to me that it is also possible to perform more advanced feature extraction on the control side and think about perceptual features on the sound engine side. Indeed, it is easy to forget that music theory is at a much higher level of abstraction than sound theory when you are stuck with reading in raw sensor signals and figuring out ways to pass all the data forward in a processing chain. From then on, I turned my attention toward "perceptual mapping." This chapter will look at my proposed Gesture Description Interchange Format (GDIF) and why it failed. But first, we will take a look at traditional musical notation and the commercial standards MIDI and MPE. After all, they, too, are formal representations of musical information based on perceptual features.

Coding Actions in Musical Scores

It may not be evident to everyone, but traditional musical scores are a form of action notation. A "note" in a musical score is a description of how a "tone" is going to be played. Figure 6.1 shows an elementary score, containing only a few notes. The beauty of musical notation is that it manages to carry a lot of information in an incredibly compact visual representation. For example, the vertical position of a note in the score tells which pitch to play. This assumes that one knows the "clef," which serves as a reference point of the score. The score in figure 6.1 is written in

Figure 6.1
A musical score contains a lot of information in a compact visual representation.

"treble clef." This means that the second line from the bottom represents the tone G. In addition, it is necessary to check whether there are any signs at the beginning of the system that indicates the "key" of the tune (here, D major or B minor). In the example, there are two ♯ written at the beginning of the score, which means that all the F and C notes should be played sharp, that is, as F♯ and C♯, respectively. This is unless there is a ♮ sign in front of a note, which would remove one of the global signs. In addition, there could also be extra ♯ or ♭ signs for single notes. It should be said that the whole system is based on a twelve-tone equal temperament scale, so the steps between notes in the score are unequal. There are whole steps between some notes (C–D, D–E, F–G, and A–B) and half steps between others (E–F and B–C).

There are also tuning differences to take into account. In the example in figure 6.1, the first note will be interpreted as the tone D. However, that D will not necessarily sound the same when played on different instruments. If played on the piano, the tone will most likely have a frequency of around 293 Hz. If you play the same D on a regular trumpet (tuned in Bb), it will sound like the C on the piano, while a D played on an alto saxophone (tuned in Eb) will sound like an F. So the instrument matters a great deal when turning notes into tones.

There are also microtuning differences between instruments. When I mentioned that a D played on the piano results in a tone with a fundamental frequency of 293 Hz, this was based on the assumption that the piano is tuned in a performance pitch of 440 Hz. The performance pitch is usually defined by the frequency of the note A above middle C. While 440 Hz is a standard tuning today, Haynes (2002) has documented that the "A" has had many different pitches throughout history. And also, today, different ensembles explore other tunings.

For musicians playing alone, the exact tuning does not matter. The most important is that the instrument is tuned relative to itself. So the instrument being tuned in A = 430 Hz or A = 450 HZ does not matter for people without perfect pitch. That said, many musicians use auto-tuners these days, and they are typically based on a frequency of A = 440 Hz. So are most commercial electro-acoustic instruments, such as digital pianos. Various digital audio workstations may also be set up to work best with a standard tuning of A = 440 Hz. Thus, new technologies have lead to a standardized tuning practice.

What is fascinating is that musical scores are both absolute and relative at the same time. The score attempts to specify an absolute ratio between notes, but these ratios are relative at many levels. The performer needs to "calculate" the right note based on knowledge about music theory. In addition, a score also contains both absolute and relative temporal information. The position of notes from left to right tells the performer about the order of tones to be played. However, as opposed to the even distribution of points on the *x* axis in data plots, a musical score does not base its timing on the notes' exact position on the page. It is a note's value—such as 𝅝 𝅗𝅥 ♩ ♪ 𝅝. 𝅗𝅥. ♩.—that tells how long it should be played. These relative durations can be made even more relative by using a 𝄐 (fermata). Scores also often have a tempo indication. This could be a relative representation, such as "Andante," which would indicate a tempo in the region of around seventy to one hundred beats per minute on a metronome. There could also be a fixed number (♩ = 80) prescribing the intended beats per minute. However, how precisely such a number is interpreted is up to the performer.

In addition to frequency and time information, a score may contain loudness information, usually specified in relative terms. It can either be through a textual description (*ppp—FFF*) or a crescendo line, as in figure 6.1. Finally, a score may contain "absolute" information about timbre by specifying an instrument. Additional "relative" timbre information can be added through descriptions of playing techniques. For example, in violin scores, the performer would play with a bow if nothing else is specified. But if "pizz." is written, the performer would play with the fingers instead.

Given the large "gap" from note to tone, much interpretation is left to the performer. I still remember one of my first composition classes, in which we were given the task of writing a short cello piece. I had made my score in notation software and had listened to the piece using the simple MIDI playback device available in the software. This had rendered the melody sufficiently well to understand how it could sound. However, hearing the piece performed by a live musician, with the beautiful cello sound right in front of me, was mind blowing. It was a compelling manifestation of the difference between notes and tones.

Musical scores have developed over many centuries to become a powerful and flexible tool for composers and musicians (Grier 2021). Many composers have used scores as their primary method to communicate musical intentions in the Western art music tradition. Music researchers have also embraced scores as a method to study musical structures and composers' intentions. This does not only include music that was notated from the start. It has also been a tradition of transcribing sounding music into written scores for later analysis, such as jazz improvisations or folk tunes. However, one should always be careful about simplifying and interpreting such "translations" from one representation to another.

Chord Notation

The note-based musical scores mentioned above are just one of several ways that musical information can be coded. Another commonly used score type is that of chord notation. Such notation is typically found in books of popular songs or jazz tunes and is even more "action based" than traditional music notation. Chord notation is based on providing only information about the chords to be played. For example, a standard jazz chord sequence can be written as Dm7, G7, C. Those familiar with this type of notation know that Dm7 includes the tones D-F-A-C, G7 consists of the tones G-B-D-F, and C has C-E-G. If this were a jazz tune, many would probably also think that the C is a short form of writing Cmaj7, including the major 7 in the chord (C-E-G-B). Someone familiar with jazz theory would immediately be able to improvise a melody if they were given a sequence like Dm7, G7, C. Thus, such a sequence is an even more compact way of representing musical information than a note-based score. It also requires even more knowledge about music theory and the particular genre in question.

While it is common to write out both the note name (D) and the chord structure (m7), the representations could be even more compact. In functional harmonic analysis, it is common to write IIm7, which would be a relative pitch representation. In the example above, II would represent the second note (D) of the scale of the fundamental note (C). In genres where there is a clear understanding of how the harmonic structures relate to each other, it may be sufficient to write the above example as II, V, I. This would be sufficient to understand which chords and tones to play. Writing notes with such a functional representation also makes it easier to transpose a tune to a different key. For example, if I would like to play the tune in G major instead of C, the II, V, I sequence would be played as Am7,

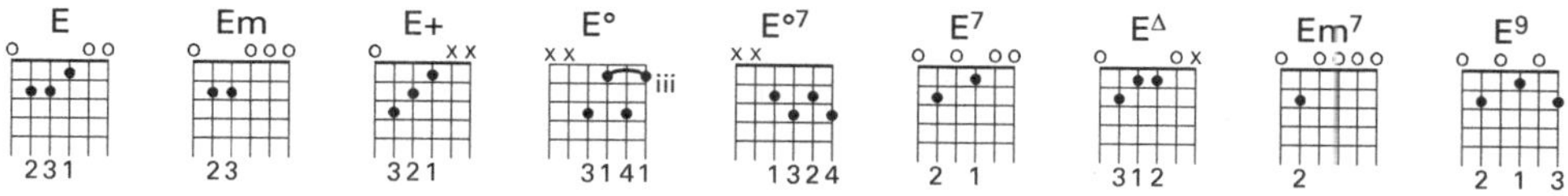

Figure 6.2
Chord notation (top) with corresponding guitar fretboard diagrams.

D7, G. Such translations between different representational systems are standard procedure for many musicians.

Some people find it challenging to remember which notes to play from chord notation. For that reason, some guitar books contain fretboard diagrams, such as shown in figure 6.2. These are visual representations of the guitar neck, indicating open strings and the frets that are necessary to hold to get the right chord. How to play them, however, is up to the performer. One can strum the guitar strings, thereby playing all tones at (almost) the same time. Or one can play the bass tone followed by the other tones. Chord notation contains no timing information beyond indicating when one should change from one chord to another. I find it interesting that various types of music notation, whether note based or chord based, are predominantly based on representing actions. In guitar chord notation, it even contains a simplified representation of a part of the instrument.

Magnusson (2019) discusses how musical scores have developed over the centuries and how this has shaped musical creation and thinking. This includes the traditional note- and chord-based scores mentioned above and various types of computer code and other representations of musical tones and ideas. The experimentation with extended techniques on acoustic instruments leads to the continuous expansion of the symbols used in scores, although many of these are not standardized. Also, compositional experimentation leads to new types of scores. While traditional, linear scores still prevail, examples of circular or fragmented scores allow for more flexible performance. The addition of new electro-acoustic instruments continues to push for new notation forms, both on paper and on screen. Some graphical music programming languages (such as PureData and Max) may contain score information in the graphical user interface itself. Other languages, such as CSound, uphold a clear separation between sound ("orchestra") and structure ("score") (Boulanger 2000). Despite these differences in representation—spanning from traditional to experimental score types—there is a need to represent information about musical tones and musical structure. In computers, this is most commonly done with the MIDI format.

The Musical Instruments Digital Interface (MIDI)

When electro-acoustic instruments, particularly various types of keyboard-based synthesizers, became popular in the 1970s and early 1980s, it became clear that there was a need to standardize the communication between devices. This led to the development of the Musical Instruments Digital Interface (MIDI) format, which the MIDI Association proposed in 1983 and which has been the de facto instrument communication standard ever since (Jungleib 1996). The MIDI standard provides a simple way of connecting controllers to sound engines by defining the physical cables used to connect devices and the protocol used for the communication. The MIDI standard does not specify mappings between controller and sound, but the General MIDI (GM) specification and Roland's General Sound (GS) extension define specific timbre sets. This makes it possible to play a MIDI file on any GM-compliant system and get a more or less similar rendition of the music.

MIDI is one of the most successful and oldest standards in the computer industry (Rothstein 1992), and one that is still going strong. That is the case even though it was met with criticism already from the start. Both Loy (1985) and Moore (1988) criticized MIDI for its low resolution, high latency, number-based messaging structure, and serial nature. While each of these weaknesses may not be that problematic in itself, the combination often led to challenges, particularly in large setups. Several of these problems have later been solved or marginalized in various ways. For example, MIDI signals were traditionally passed over five-pin DIN MIDI cables, but it is now more commonly transmitted through USB cables. The growing processing speed has also helped in reducing latency problems. So for many use cases, MIDI is good enough for the task at hand.

In the context of this book, my main interest in MIDI is its conceptual design, particularly the use of "notes" as the central organizing unit. MIDI communication is built around sending "note-on" and "note-off" messages using a twelve-tone equal temperament system to denote the pitch values. This fits well with the traditional, Western music notation systems described above. It also works well with the discrete nature of keyboard-style controllers, modeled after the piano's sound-producing actions. In fact, for such impulsive actions and sounds, MIDI provides a highly efficient representation. However, such an impulsive-based representation is challenging for continuous controllers and sustained sound engine types. For example, representing a continuous pitch slide with MIDI requires

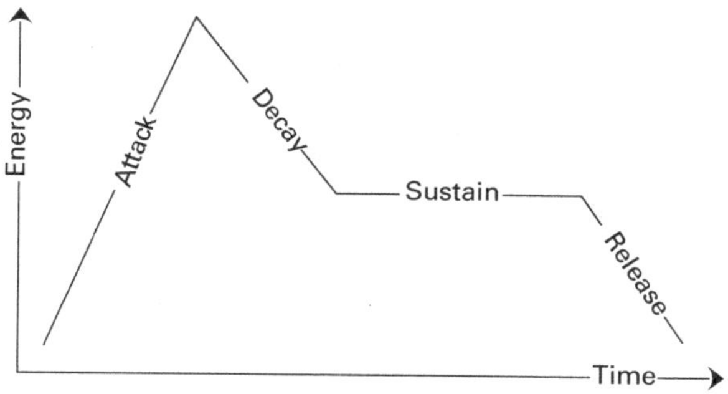

Figure 6.3
A sketch of an ADSR envelope, which is commonly used to shape the tone's amplitude in electro-acoustic instruments.

the use of a series of note-on, note-off, and pitch-bend messages. Working with microintervals is equally problematic, as one would need to send tempered pitch messages together with a pitch-bend message denoting the "detuning" of each note.

The MIDI standard encourages impulsive-like sound control. One can, therefore, only wonder whether the MIDI standard itself is a reason for the proliferation of keyboard-based and "knobs-and-sliders"-type controllers. Since only the tone's attack can be controlled directly—with information about its pitch and the initial velocity—the rest of the tone has to be controlled programmatically. This can be done through the use of "envelopes," such as the attack-decay-sustain-release (ADSR) envelope (figure 6.3). Many synthesizers allow modifying the amplitude and duration of each part of such an envelope using separate controllers. This can sometimes be done during a performance, although it would mean that the performer moves one or two hands away from the keys. It is more common to adjust such settings between songs.

Low-frequency oscillators (LFOs) are another group of temporal control messages that are common to use in synthesizers. These are time functions designed to continuously change the shape of sounds and their timbral/textural features. An LFO can be anything from a low-frequency sine tone to complex multifrequency signals. In some cases, they may also include rhythmic, melodic, or harmonic elements. As we will discuss later, such processes are part of the change from "sound makers" to "music makers." For the performer, such long semiautomatic processes may feel like the instrument plays "itself." In some cases, the system may even be designed so that the performer knows when to release the key. Lossius

(2007, 77) reflects on how such procedural processes may lead to the instrument controlling the performer rather than the other way around. Most synthesizers allow for programming both envelopes and LFOs in various ways, and parts of the creative process may be spent on making "presets" for a performance. In some cases, entire pieces can be composed by solely setting up synthesizer presets. Synthesizer manufacturers have for a long time understood the importance of providing high-quality presets with their devices. Users appreciate products that have a varied set of presets available. There is even a market for selling and buying presets that extend synthesizers' possibilities.

MIDI Polyphonic Expression

After many years of discussion about how to extend the MDI standard, the MIDI Polyphonic Expression (MPE) extension was finally agreed on in 2018 (Association 2018) followed by the MIDI 2.0 specification two years later (Association 2020). These extensions came out of the need for interoperability of new keyboard-based instruments with continuous control (which we will look more at in chapter 10). The new standards are implemented conservatively, meaning that they are backward compatible with the original MIDI specification.

The MPE standard is implemented using MIDI channel 1 for sending global messages. The remaining channels transmit notes and "expressive" data, including note-on velocity, pitch bend, channel pressure (aftertouch), brightness, and note-off velocity, on a note-per-channel basis. This makes it possible to send richer messages than what is possible with traditional MIDI streams while still staying within the MIDI ecosystem. A similar approach is taken in MIDI 2.0, but here with a broader implementation based around a MIDI Capability Inquiry (MIDI-CI) system (Association 2020b).

MPE and MIDI 2.0 solve some of the problems with the original MIDI standard, but not all. The standards are still focused on keyboard-based control and the concept of note-on and note-off messaging. Even though it now supports continuous control, the MIDI standard is still not a generic solution for object-based sonic interaction.

Open Sound Control

It is easy to spot problems with the MIDI standard, but it is challenging to develop alternative solutions. Throughout the years, there have been several failed attempts. The most successful alternative is Open Sound Control (OSC), which was proposed by a team of researchers at the Center for New Music and Audio Technologies (CNMAT) at the University of California, Berkeley, in the 1990s (Wright et al. 1997). Over the years, OSC has become the de facto communication standard in the music technology research community (Wright and Freed 2017), and it is also used in some commercial products.

OSC was designed as a general communication protocol and can be used over many different cable types, including Ethernet, USB, or even MIDI cables. It adopts a URL-style messaging format, such as

```
/voices/3/freq 400
```

which refers to sending the message 400 to the module `freq` inside a node 3 that is a child of the top-level node `voices`. As opposed to MIDI's serial messaging nature, OSC supports pattern matching. This means that it is possible to send messages like

```
/voices/*/freq 400
```

to control the frequencies of several voice modules in parallel. OSC's text-based labeling structure allows for sending meaningful messages between devices. This makes the protocol more human friendly than the number-based approach used in MIDI messages. Still, OSC has been criticized for being too lightweight (Dannenberg and Chi 2016). It relies on IP addresses and port numbers without handling human-friendly device names.

As the name implies, OSC is an open standard in which few limitations are forced upon the user. This openness is both a strength and a limitation. MIDI is based on the "note" as a core sonic/musical entity, and any MIDI-compliant interface could be connected to any sound engine. OSC allows the user to create any message. That may be liberating at first, but the result is that OSC-based devices and software cannot communicate directly without first setting up specific mappings between them. Already from the start, it was suggested that working toward standard guidelines and creating uniform namespaces was important (Wright et al. 2001). However, except for a few examples—of which the TUIO standard for table-based tangible user interfaces is the most prominent (Kaltenbrunner et al. 2005)—such standardization has not happened. That may be one reason that OSC has gained the most traction in the research community, in which people are more inclined to develop namespaces themselves.

Gesture Description Interchange Format

My interest in developing a higher-level mapping system for control interfaces and sound engines led to the proposal of the Gesture Description Interchange Format (GDIF) (Jensenius et al. 2006). This was an ambitious project, aiming to create more human-friendly mappings from action to sound. The dream was to have a system that would allow for mapping between multiple data streams, ranging from low-level sensor data to higher-level control features, such as sketched in figure 6.4.

Inspired by the ESCHER system of Wanderley (2001), GDIF aimed at using "intermediary" mapping layers. Instead of mapping directly from technical action parameters to technical sound parameters, the user would map between perceptually meaningful parameters (figure 6.5). For example, mapping the position coordinates of a joystick onto the center frequency and gain of a filter may seem logical to people with music technological experience. However, such a technical mapping would probably not be intuitive for most others. A more ecological approach would be to create mappings based on the action used ("move the right hand to the right") and the perceptual quality of the sound ("change the brightness and loudness of the sound").

The idea was to base mappings on meaningful parameters and allow for interchanging controllers and sound engines. So the perceptual mappings

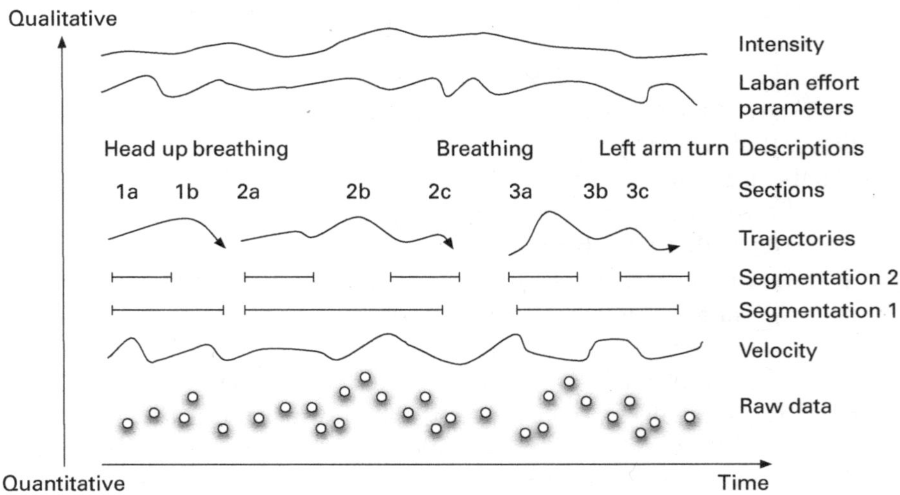

Figure 6.4
Sketch of different levels of data that could be streamed with GDIF.

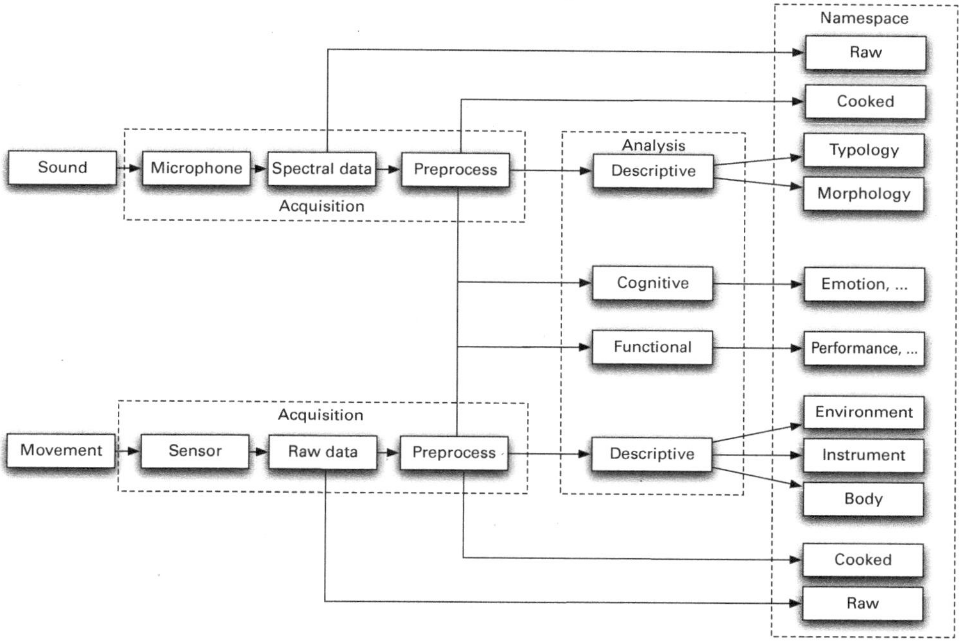

Figure 6.5
Sketch of the multilayered namespace proposed for GDIF.

would remain while the technical mappings would change. For example, if I want to use a lateral hand action to control a sound, it should not matter what type of motion capture system I use to track information about the action. A camera-based or sensor-based system will have different numerical representations of the motion in question. However, as long as these are first mapped into a meaningful motion-based mapping layer, this layer can be used for mapping onto perceptually meaningful sound features. On the sound side, one could imagine a similar type of intermediate mapping layer between perceptually meaningful sound features and the technical features of the particular sound engine.

I still think that the conceptual idea of GDIF is good, but it is challenging to implement. I created some mapping examples that worked well (Jensenius 2007). However, scaling up to a general solution showed how challenging it is to handle all the nuances and complexities of musical instruments. It would certainly be possible to develop coarse generic mappings, but I doubt that the dream of creating a genuinely generic mapping system could ever work. If so, machine learning–based approaches have proven to be more successful in achieving such a goal (Fiebrink et al. 2011). Then the user can choose to train the system based on perceptually meaningful actions and sounds. Given the "black box"–nature of such models, the user would not get a meaningful representation of the mapping. However, that may not necessarily be a problem if the mapping works as expected.

I have used machine learning in some of my instruments, but many of the instruments I will present in later chapters are based on basic technical mappings. After working on more sophisticated mapping systems for some years, I discovered that creating coupled mappings between a few technical parameters can work quite well. This also gives some "resistance" to an instrument that may spark creativity. However, from a conceptual perspective, I still think it is an interesting exercise to reflect on the possibility of creating perceptually based mappings.

III

Interaction

7

Action–Sound Couplings

"Toot Whistle Plunk and Boom." That is how the history of Western musical instruments was introduced in Disney's 1953 cartoon of the same name. The four words are onomatopoeia for the sounds of brass, woodwinds, strings, and percussion instruments, respectively. I often show the cartoon when lecturing about instruments. It is funny and yet remarkably illustrative of what I call "action–sound couplings." I call these couplings because there is always a sonic result when you act on a physical object, as sketched in figure 7.1. If you interact lightly, the sound may only barely be audible. But it will be there if you listen carefully, and certainly if you attach a contact microphone to the object in question.

Sound-Source Perception

An action–sound coupling can be described in terms of the acoustic properties of the objects involved and the (bio)mechanics of the sound-producing action. Given the same physical objects and actions, the resultant sound will always be the same. This is also reflected in the way we perceive the interaction. Our life-long experience with various objects makes us able to predict the sonic result of many actions on those objects. This means that we can predict the resulting sound of an interaction even before it is heard.

We can verify our mental capacity for predicting the outcome of an action–sound coupling in a thought experiment. Think back at the example of the glass falling toward the floor. While still in flight, you will imagine both the visual and auditory results of the glass breaking to pieces when it hits the floor. Knowing that the sound will be unpleasant, you may even try to cover your ears to protect your hearing. This example demonstrates that we immediately understand the sonic properties of the objects involved, the actions on those objects, and the resultant sounds. This

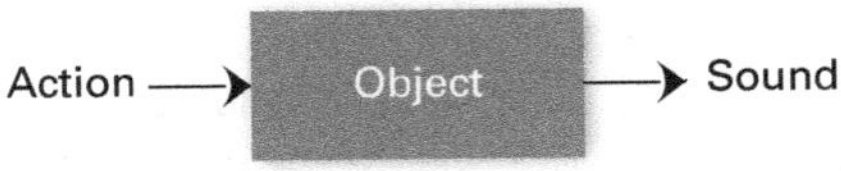

Figure 7.1
Acting on a physical object will lead to sound production.

knowledge can be used to estimate both the timbral qualities and loudness of the sound to emerge based solely on the glass' visual information and its trajectory in the fall. Such tacit "psychomechanical" knowledge guides us in also telling the approximate size of the glass, the surface it hit, the distance it fell, whether it was dropped or thrown, and so on (McAdams 2000).

Let us turn the thought experiment around and start by imagining that you only "hear" the sound of a glass breaking to pieces. Even though only the audition is activated, you will be able to tell much about the objects and actions involved. Researchers in sound-source perception have uncovered our remarkable ability to perceive the qualities of the materials and actions associated with sounds only from listening (Gaver 1993a, 1993b). This ability to identify the properties of objects based solely on sound seems to be surprisingly reliable and accurate. It is robust during attentive listening and in the everyday perception of impact sounds and sources (Rocchesso and Fontana 2003). This is logical from an ecological perspective, considering that characterizing and identifying sounds is essential in the interaction with objects in the environment. This means that we can recognize qualities of specific types of sounds, such as bouncing and breaking bottles (Warren and Verbrugge 1984), clapping hands (Repp 1987), mallets striking pans (Freed 1990), the amount of liquid in a cylinder (Cabe and Pittenger 2000), the direction of a sound source in three-dimensional space (Neuhoff 2001), the length and hollowness of objects (Carello et al. 1998; Lutfi 2001), their shape (Kunkler-Peck and Turvey 2000), the material categories of struck objects of variable sizes (Giordano and McAdams 2006), and so on. I find these and similar studies fascinating because they attest to the richness of our auditory system. They also support my main argument that action–sound couplings are essential in our interactions in the world.

Excitation and Resonance

The physics of action–sound couplings are generally well understood. So are the basic psychoacoustic principles. Then there are more open

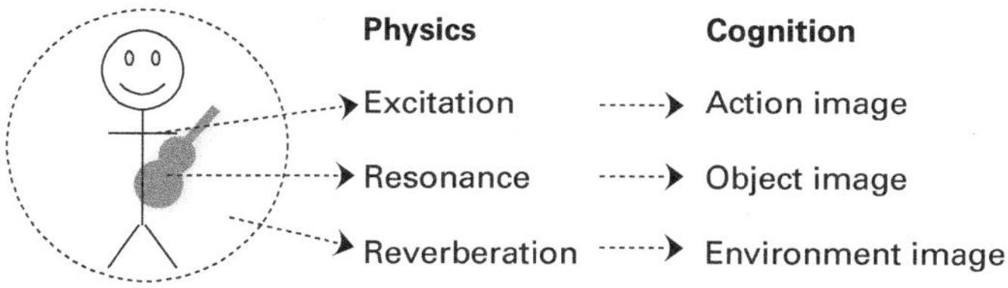

Figure 7.2
The mental imagery of sounds can be broken down into three components: images of the actions, the object(s), and the environment.

questions about how they are perceived by our (embodied) cognitive system. Godøy (2001) has suggested that our cognition of sound-producing actions may be based on "images" of sonic features. Such mental images should be understood as "quasi-perceptual" experiences that resemble real perceptual experiences but with the absence of external stimuli (Thomas 2007). Even though the word "imagery" may draw attention toward the visual, all modalities are included when "mental imagery" is used. This includes both "seeing in the mind's eye" and "hearing in the head." Therefore, the term "musical imagery" can be understood as the mental capacity for imagining musical sound in the absence of a directly audible sound source. Given the close connections between action and sound, such a "hearing" of sound in your mind would also mean that you "see" the accompanying sound-producing action.

In his elaboration on the mental imagery related to musical sound, Godøy (2001) further suggests that such imagery can be divided into "motor images" and "material images." Here the motor images are mental representations of the excitation of an object. The material images represent the resonances of the sounding objects, which indicate the properties of the objects. I would argue that we also hear one extra layer of resonance: the environment. If you imagine a drum stroke, for example, you will be able to hear the drum and its properties (size, shape, materials) and the object that hit the drum and its properties (hand, stick, mallet). You will also most likely hear the room within which the sound appeared and its properties (small/large, dry/wet, and so on). These different layers are summarized in figure 7.2. The sounds of the environment are studied within room acoustics— a different field from instrument acoustics—and have often been neglected in many discussions about instruments. While this makes sense in many situations, I have found that it is necessary to take this layer into account to understand the differences between acoustic and electro-acoustic instruments.

In a further development on his thinking on musical imagery, Godøy (2006) suggests that our cognition of sound objects is based on mentally "tracing" features of the sound while listening to, or even when only imagining, music. This was the idea behind the sound-tracing studies described in chapter 5. These studies showed no one-to-one relationship between the physical properties of the involved actions and their imagined sonic counterparts. We do not, for example, see a guitar when hearing a guitar-like sound, but we get an embodied sensation of strings and a vibrating body. In my extended model, we also get an embodied feeling of the physical or virtual space where the vibrating body is located.

The Sound Object

The model of musical imagery put forward by Godøy (2001) builds on the thinking of Schaeffer (1967) on "reduced" listening. This is a particular type of listening that is focused on the sound "itself." At the core of Schaeffer's theory is the identification of "sound objects," fragments of sound perceived as one coherent unit. Sound objects can last up to a few seconds but are often shorter. The main point is that they are perceived holistically as an "object" with a shape. It may seem paradoxical that I use the concept of a sound object in discussions of its sound-producing action. After all, Schaeffer explicitly wanted to listen to sounds without considering the sound source. The aim was to use sound material in his compositional practice. Schaeffer did this by cutting and splicing sound recordings so that the original sound source was unidentifiable. He also discovered that much of the source information was lost when cutting off the beginning of the sound, the attack part. This made it possible to listen to the timbral qualities of the sound and not get stuck with mental imagery related to the objects and actions involved in sound production.

Several composers and music researchers have extended Schaeffer's thinking over the years. Smalley (1997) has developed his "spectromorphology," which refers to the interaction between sound spectra (the "spectro-") and how they are shaped through time (the "-morphology"). Thoresen and Hedman (2015) have developed a complete visual language for spectromorphological analysis. Common to their—and Schaeffer's—thinking is that the sound should be considered without reference to its source. Smalley (1997) refers to this as getting away from "technological listening," that is, listening for the techniques of sound production rather than the sound itself. In acousmatic music, the loudspeakers are the physical sources of the audible sound. The sound's identity, however, is not visible. Still, there may be cases in which the listener tries to (willingly or unwillingly) locate the source of a sound, what Smalley (1997, 110) refers to as "source bonding":

> . . . the *natural* tendency to relate sounds to supposed sources and causes, and to relate sounds to each other because they appear to have shared or associated origins.

Godøy (2006) argues that, despite the seemingly "disembodied" nature of Schaeffer's thinking, there is no conflict between the principles of reduced listening and those of embodied cognition. Instead, there are several similarities between studying sound objects and their sound-producing actions from a phenomenological perspective. This is the case also for my thinking about the cognition of action–sound couplings. When we hear a guitar-like sound, we get a sensation of an impulsive-like attack. Such a pluck on a guitar will result in quite different imagery than when hearing the bowing on a cello, even though both are string instruments with similarly shaped bodies. It is also interesting to consider the difference between hearing plucking versus bowing on a string instrument like the cello. The sounds would be perceived differently, but the sound source (the cello) would probably be heard. This is an example of how excitation and resonance play together in the cognition of sound objects. It is important to remember that this model is bidirectional. Hearing a sound object will evoke the mental image of a sound-producing action. Similarly, seeing a sound-producing action will evoke a mental image of the corresponding sound object.

Developing this thinking one step further, Launay (2015) argues that listening to music is an inherently social experience, even when it happens alone. He proposes a model in which the detection of human agency is at the core of the musical experience. This is based on learning relationships between sound, action, and agency. Launay et al. (2016) followed up with an empirical study in which nonmusicians were asked to quickly learn new associations between sounds and observed actions without performing those actions themselves. The researchers measured the motor-evoked potentials (MEPs) in the participants' right hands and used single transcranial magnetic stimulation (TMS) pulses to trigger motor activity. The results showed that observers could quickly learn to associate sound with action. This indicates that it is not necessary to have played an instrument to experience motor resonance when hearing that instrument.

Most of the literature that I build my action–sound theory on has explicitly tried to get away from understanding the sources of sounds, whether these sources are instruments or other types of sound-making objects. However, as mentioned above, there is not necessarily a conflict between understanding (and appreciating) the qualities of sounds while at the same time reflecting on how they are produced. In fact, knowing more about the perception of acoustic sounds may help create better and more interesting electro-acoustic sounds. Therefore, let us move from the general thinking about action–sound couplings to some more music-specific examples.

Action–Sound Couplings in Instruments

As mentioned in chapter 3, Paine (2015) has proposed the term "techno-somatic dimension" to describe what is happening between the physical instrument and the human body. He argues that the techno-somatic dimension is the "somatic gestalt" formed between the physical instrument and the human body's intentional actions. This somatic gestalt represents the instrument's cumulative experience, enacted through technique and experienced through somatosensory and listening phenomena. This relates to my thinking about the action–sound palette of an instrument (or any other sound-producing object).

Paine further argues that many new designers of new instruments/interfaces have forgotten the techno-somatic dimension. The most significant problem with general-purpose interfaces, he argues, is that they do not offer an idiomatic playing style. Many standard acoustic instruments have a clearly defined playing technique, which helps develop performance skills, pieces, and educational material. To make successful new instruments, Paine suggests putting the techno-somatic dimension on top of the design priority list. He also argues that building on existing instrument designs makes it possible to draw on the transferable skills of "neighboring" instruments. I partly agree with this but will later reflect on some challenges assuming that skills are transferable between devices.

The generic, one-directional action–object–sound chain presented in figure 7.1 does not reveal the *inter*action going on in musical performance. Typically, there is a continuous loop of actions exerted on the instrument by the performer and *re*actions from the instrument to the performer, as sketched in figure 7.3. The actions excite the instrument, which leads to vibrations in the instrument body. These vibrations lead to audible sound, and they also lead to haptic feedback to the performer.

Most instruments are constructed from multiple parts, and many may also be played with one or more tools. For example, the two most essential parts of a violin-like instrument are the strings (the "vibrator") and the

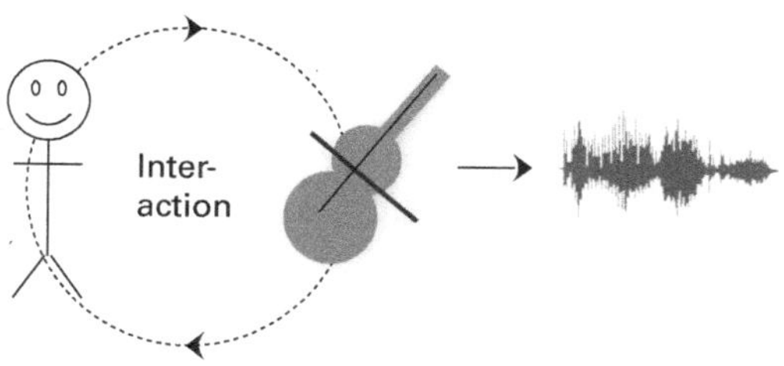

Figure 7.3
An instrument can be seen as a mediator between action and sound, including a constant feedback loop between instrument and performer.

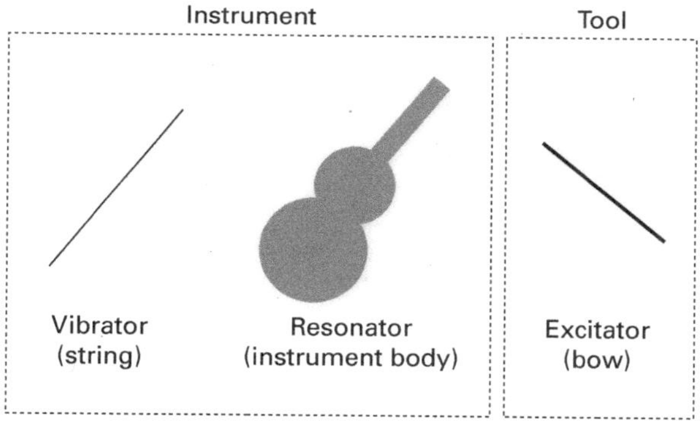

Figure 7.4
A string instrument makes sound with a vibrator, an excitator, and a resonator.

instrument body (the "resonator"). To produce sound, we also need a bow (the "excitator"), as sketched in figure 7.4. Many percussion instruments are tool based, while brass and wind instruments are not. In the latter, the excitation happens at the meeting point between lips and mouthpiece. Still, all of these instruments rely on an instrument body that acts as a resonator.

I will not go into further details of instrument acoustics here, but interested readers can check out the excellent overview in Rossing et al. (2002). Here we will keep the discussion at a system level, from which it can be argued that the result of an interaction with an instrument ultimately boils down to the physical interaction between the objects involved, which again will be based on the following parameters:

- The instrument body and its vibrating parts
- The tool(s) used to play the instrument

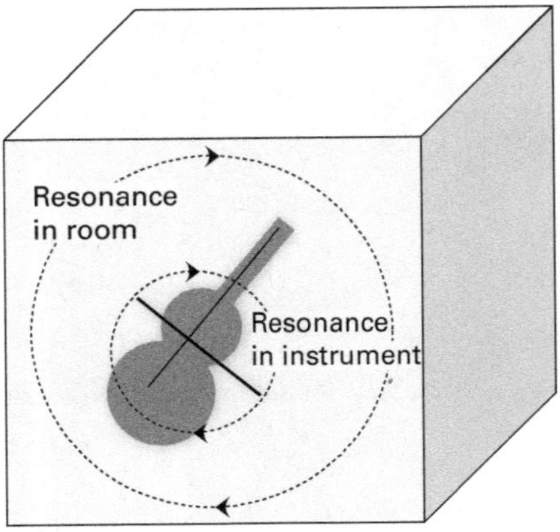

Figure 7.5
A sketch of the two layers of resonance, one in the instrument body and one in the room.

- The sound-producing action
- The environment within which the interaction happens

As mentioned in the discussion about mental imagery, the environment is often forgotten in conversations about instruments. Even though most musicians have preferences for particular environments, most would probably think about their instrument as moving from space to space. However, as we shall see later, there are also cases in which the environment is crucial for how an instrument is performed and perceived. So I am in favor of including the room that the instrument is played within as part of a general instrument model (figure 7.5).

Action–Sound Palette

As discussed in chapter 2, an object can have multiple affordances, that is, invite different types of usage. It may also have numerous sonic affordances. These sonic affordances are related to the size, shape, materials used, and construction of the sound-producing objects involved in an interaction. The sonic affordances help set up an action–sound palette of possible couplings between actions and sounds. So an action–sound palette can be seen as our combined knowledge of the properties of sounding objects and possible sound-producing actions.

Action–sound palettes are probably deeply rooted in our cognitive system and is something we activate continuously in our interactions in the world. This helps us avoid making unnecessary loud sounds when we move around. Other times, we may want to make sound to let others know that we are coming. The same system is activated in musical practice. For example, think about how you can make music with a wine glass. You can use it as a percussion instrument by tapping on the side with your finger. Tapping with the fingertip will result in a different sound than using the nail. Both would be examples of impulsive sound-producing actions, albeit with different sonic results. You could also wet your finger and move it in a circle on the top of the glass to produce sound. Furthermore, the pitch of the sound could be modified by pouring water into the glass. These possible sonic affordances (and more) would together constitute our understanding of the action–sound palette of the glass.

As we shall see later, the idea of a constrained action–sound palette does not work equally well for action–sound mappings in electro-acoustic instruments. In theory, any combination of action and sound can be designed and created digitally. This makes it difficult to set up a similar set of expectations for an action–sound mapping than you would for an action–sound coupling. When you touch the key on a keyboard-based synthesizer, it is impossible to know what type of sound you will get.

Action–Sound Separation

The concept of an action–sound palette helps define the possible sonic outcomes of an interaction. As we move toward discussions of musicking machines, it may also be relevant to have a conceptual apparatus for talking about the body's involvement in the sound production. Keislar (2009) uses the term "disjunction" to explain the increasing disconnection between action and sound throughout the development of instruments. To create a more fine-grained terminology of such disjunctions, I will introduce the term "action–sound separation" to describe how close (or far) the body is related to the sound-producing object. This is inspired by the Hungarian composer Bela Bartók's organization of instruments according to their closeness to the human body. In a 1937 essay on mechanical instruments, Bartók (1976, 289–290) suggested a continuum from the human voice, followed by wind instruments, bowed string instruments, plucked string instruments, pianos, organs, barrel organs, player pianos, gramophones, and, finally, the radio.

Inspired by Bartók's continuum, Thelle (2010) suggested five levels of action–sound separation in instruments: incorporated, direct, mechanical, analog, electronic, and digital. Enders (2017) has identified ten developmental stages from the human voice to virtual musical instruments:

instrumentalization, mechanization, automatization, electronification, modularization, digitalization, virtualization, globalization, informatization/artificial intelligence, and hybridization.

There are compelling arguments for thinking about a continuum from the most embodied acoustic instruments to the purely digital ones. However, based on my techno-cognitive reasoning, I believe there are even better arguments for thinking about acoustic and electro-acoustic instruments as two separate categories. After all, acoustic instruments are based on action–sound couplings governed by the laws of physics, while electro-acoustic instruments are based on designed action–sound mappings. In my current model, I have, therefore, decided to keep them separate.

We will start by considering the action–sound separation of acoustic instruments and return to electro-acoustic instruments in chapter 8. I propose the following elements of a continuum describing the action–sound couplings in acoustic instruments:

- **Embodied:** Sound is produced with(in) the human body, such as the voice or clapping.
- **Tactile:** The human body is in direct contact with resonating objects, such as in the guitar or flute.
- **Tool:** There is a simple tool between the body and the instrument, such as in the sticks used in percussion or bowing in string instruments.
- **Mechanical:** Two or more mechanical elements (levers, pulleys, and so on) are involved in the excitation, such as in the piano.
- **Automatic:** There is no physical contact between the body and instrument during excitation, such as in a self-playing piano or music box.
- **Conceptual:** There is no physical instrument to interact with; the instrument and its sound production are only happening within your (embodied) mind.

The idea of the continuum is to reflect on how the interaction goes from being wholly embodied to completely disembodied, as sketched in figure 7.6. The traditional organological definitions presented in chapter 3 are primarily object based. However, one could argue that the Hornbostel Sachs system also contains information about how the objects can be played. My model, on the other hand, is entirely (inter)action based. While this also has its limitations, one immediate benefit is that the voice can be easily included in the model. This makes sense from a performance and perception perspective. After all, the voice is probably the most commonly used musical instrument.

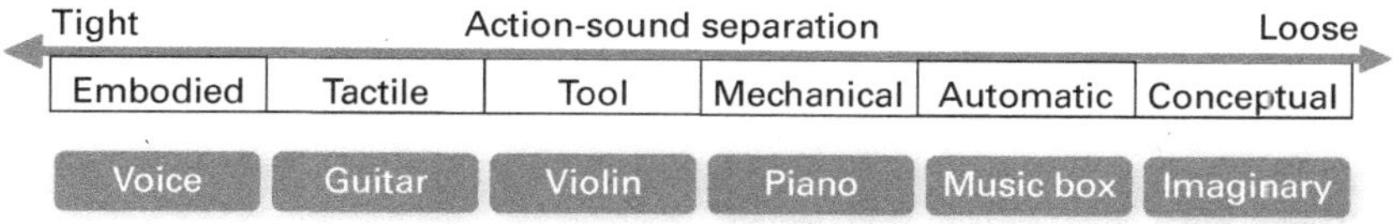

Figure 7.6
A sketch of how we may think of a continuum with an increased level of action–sound separation.

The action–sound separation presented here should be seen only as an indication of different functions. In practice, however, there are often many possible combinations. For example, string players may be in direct contact with their instrument using the left hand while playing it with a tool (a pick or a bow) with the right hand. We may also find apparent differences between excitation and modification actions regarding the level of embodiment. Therefore, it may be relevant to develop a two-dimensional continuum, to include both excitation and modification actions (figure 7.7). In the following, I will go through each of the main categories and consider how different instruments can be placed within the system.

Embodied Instruments

As should be clear by now, I favor thinking about the body itself as an instrument. The body can create lots of sounds on its own, many of which are used in musical contexts. The voice is an obvious example of an embodied instrument. It may be argued that the body was not made for musical purposes. However, it has also been suggested that human musical activity has its origins in vocalization (Mithen 2006). The body can also produce many nonvocal sounds, both intentional and unintentional. Hand clapping is an example of intentional nonvocal body sound. There are numerous unintentional body sounds, of which several have been explored in experimental performances. I still vividly remember a performance in the Yerba Buena Gardens in San Francisco of the piece *The Sound of Naked Men* by Miya Masaoka. Here the sounds of the bodies were amplified using stethoscopes placed on the bodies of the performers. The instrument in this performance was more of a hybrid nature, with amplification of the acoustic sounds from the performers' bodies. The same is the case in Marco Donnarumma's works exploring muscle sounds in music performance (Donnarumma 2016). He also works electroacoustically, relying on heavy signal processing of the original acoustic

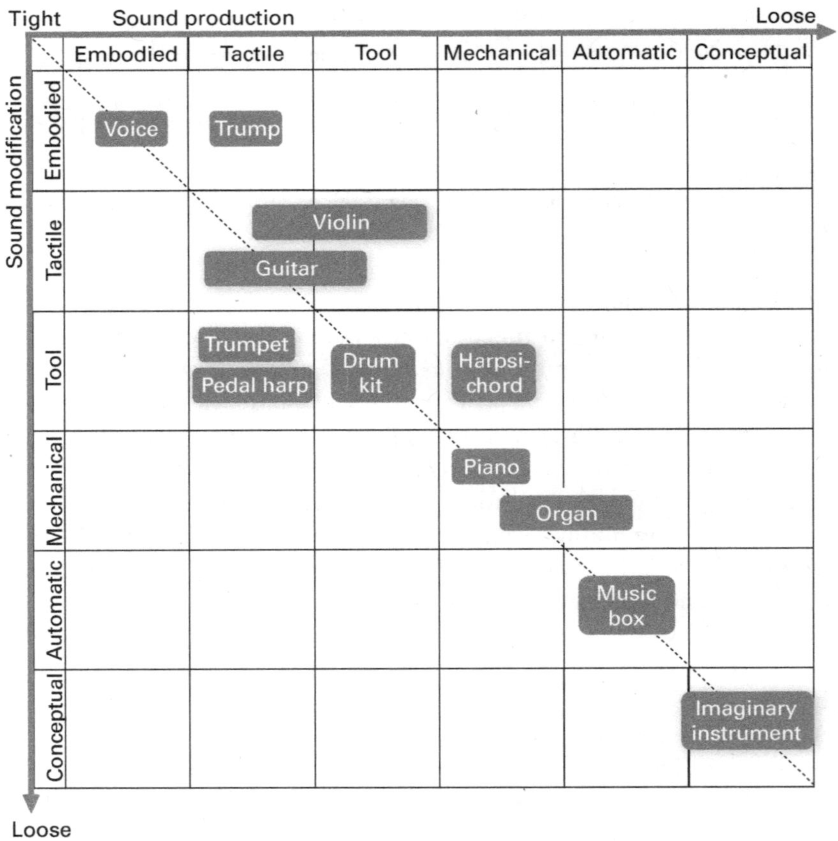

Figure 7.7
Action–sound separation in two dimensions: excitation and modification of sound, from tight to loose in both dimensions.

sound signals. Still, the original sound source is the acoustic sound of the body. We will return to further discussions of such hybrid instruments in chapter 9.

Tactile Instruments

The tactile category includes instruments in which the performer's body directly touches the sound-producing element. A hand drum is an instrument in which the performer both excites and modifies the sound directly with the hand. In a recorder, the performer excites the instrument by blowing directly into the mouthpiece while the fingers are in direct contact with the vibrating sound when closing the holes. When played with the fingers, the guitar is another example of a tactile instrument. On the other hand, if one plays on the strings with a plectrum, it would be classified as tool-based sound production. However, the left hand would still be in direct contact with the guitar. So a plectrum-based guitar performance would be classified as tool based on the sound production axis and tactile on the sound modification axis.

There are several other borderline cases within the tactile category, and which may be placed outside the diagonal in the system sketched in figure 7.7. As discussed earlier, the trump (figure 3.4) is an instrument in which the sound production is tactile since the performer is in direct contact with the instrument's vibrating element. However, sound modification is primarily happening in the performer's mouth and could be classified as embodied. It could be argued that the same is true for woodwind instruments, such as a saxophone. Here, the mouth's cavity is an integral part of both the creation and modification of sound. The creation of saxophone sound happens in the reed and is amplified in the instrument's body. The difference between a saxophone and a trump is that the trump does not have any "amplifier" built in. Instead, it relies on the creation of resonance inside the mouth. So I would still argue that there is a difference between a trump and various wind instruments with built-in resonators. This also holds for brass instruments. Here the performer is producing audible sound with the lips before it enters the mouthpiece. So the excitation of a brass instrument could be considered partly embodied—the modification, too, perhaps, considering that it is possible to shape the sound with the lips. Changing tones can also be done by adjusting the pressure of the lips.

Tool-Based Instruments

The next level of separation between action and sound is when there is some kind of tool between the performer and the sound-producing element. A simple example is that of a drum played with sticks. Here both the excitation and modification are separate from the performer's direct touch (and feel). If we compare this to the hand drum included as a tactile instrument, the performance style defines how we characterize the instrument. If the performer plays the drum with the hands, it is classified as a tactile instrument. When playing with sticks, it is tool based. While this may seem strange at first, it makes more sense when thinking about how the instrument is perceived. Think about the difference between the sound of a drum played with hands and with sticks? The tactile and haptic experiences are entirely different for the performer, which will also influence the perceiver. Even if one were to only listen to (or only see) a performance of hand-played versus stick-played drum music, one would immediately hear the difference and set up different mental imagery.

Mechanical Instruments

Instruments in this category rely not only on a single tool, but also some kind of mechanical system. The piano can be thought of as the prototypical mechanical instrument with its complex hammer system. To simplify what happens, we can say that the performer hits the key, which hits the hammer, which hits the string. As such, the performer only has indirect control of the final sounding result. It is interesting to read the reflection of Bartók (1976, 289) about the piano as a percussion instrument:

> It is apparent that we should define mechanical music—in the wider sense of the term—as a music in whose creation not only the human body but also some kind of machine is involved. We are accustomed to define the machine, in the everyday sense of the term, as a rather complex construction which serves the purpose of energy transfer. But in the course of our high school studies in physics we have met with simple machines too, such as levers and pulleys. Therefore, if a lever is a machine, then any music is also mechanized music if its origin derives from the use of levers in conjunction with the human body. Having made that statement, of which instrument are we reminded? Undoubtedly the piano, since man's finger on that instrument makes use of a series of levers for energy transfer.

The piano is but one of many different types of keyboard instruments. The clavichord is one of the "simplest" keyboard instruments, with a more straightforward mechanical construction than its siblings: the harpsichord, virginal, and spinet. In figure 7.7, the clavichord is categorized as mechanical on the excitation axis and as tool based on the modification axis. This is debatable but is based on the fact that it is possible to continuously control the string on the clavichord after it has been excited. This makes it possible to modify the sound as if one were playing it directly with a tool.

The organ is another borderline case of mechanical instruments. A typical pump organ would be categorized as having both mechanical excitation and modification. However, how should one think of the stops and the ability to play on multiple pipes simultaneously? Here the organ kickstarts the era of modern-day synthesizers in that it can produce more complex tones through advanced routing of the sound. As such, it may be closer to an automatic than mechanical instrument. When it comes to excitation, a church organ needs a massive airflow to create sound. This airflow has been produced in several ways throughout the centuries

(Bush and Kassel 2004). There are reports of organs based on the kinetic energy of falling water as early as the Roman era, and some of the pump organs can store air in large bags. Then the instrument could almost "play itself" as an automatic instrument. However, most pre-electric organs rely on manual pumping. Some are based on foot pedaling by the same person playing the keys. In the large church organs, there often had to be multiple people involved in creating the airflow. This is an early example of a multiperformer instrument, according to my definition. Most likely, only the person sitting at the keys would be called a musician, but I would argue that the people pumping air should also get credit for their musicianship.

Automatic Instruments

The automatic category includes instruments that can play without the direct energy transfer of a human performer in one way or another. This includes everything from small music boxes to self-playing pianos. There are also examples of traditional instruments equipped with self-playing mechanisms operated by a human performer, such as the pianola. I will not go into detail about all the different types of automatic instruments here, but interested readers can see Bowers (1972) for an extensive overview.

Music boxes, such as the ones in figure 7.8, may be seen as forerunners of today's musical reproduction systems. While they differ widely in construction, what music boxes have in common is that they are music makers; they embed the musical score. In some cases, the scores are stored as holes on paper rolls. Other times they are engraved into holes on metal plates or added as spikes on metal cylinders. These can then be played back by the instrument, often with a high level of temporal precision and with dynamic articulation. Some music boxes can be controlled directly by a human performer, while others are self-playing. This is achieved through the preservation of energy in the instrument. Some self-playing instruments use airbags, but many rely on spring systems. These are typically wound up with a crank, after which they can play on their own.

Some question whether automatic instruments should be considered "real" instruments. Yes is my answer. To reflect a little more on the question, let us consider another automatic instrument: the pianola. Many pianolas have a standard keyboard and can be played as a mechanical instrument. The only difference from an ordinary piano is that they can also play independently. In some pianolas, the energy input is done in

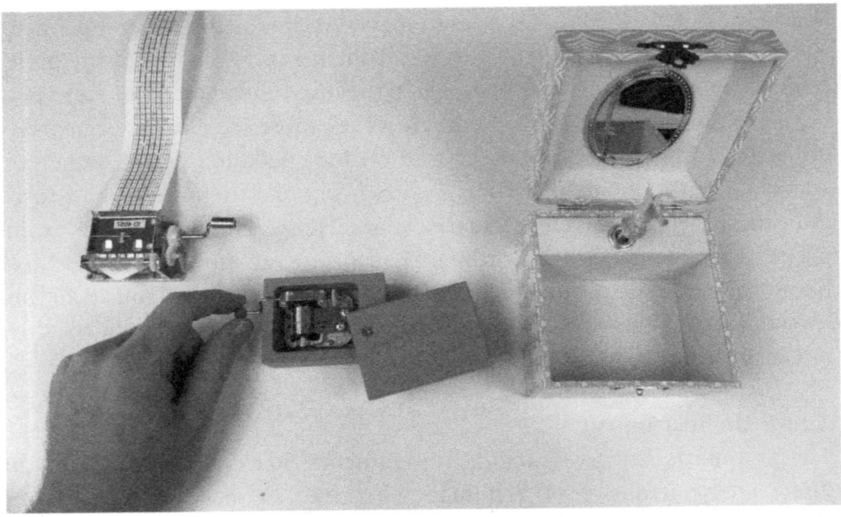

Figure 7.8
Examples of a regular music box with a small resonance chamber (middle), a customizable music box with interchangeable paper scores (left), and a self-playing music box with a crank on the back (right).

real time using a crank or foot pedals. In that case, the sound production relies on a human performing in time and in real time. The performer is in direct and continuous control of the playback speed. This is similar to many small music boxes that are played by moving the crank. The speed of the music is directly related to the speed with which one moves the crank.

If we consider that the pianola—or a directly controlled music box—is an instrument, then people using them should also be thought of as musicians. After all, a musician is someone that plays a musical instrument. Some people may object and say that playing on an automatic instrument is not a "proper" performance. However, that would be a value judgment and not connected to the fact that a person is producing music with an instrument. Such an argumentation may come from the idea that it is "too easy" to play an automatic instrument. Only turning a crank breaks with the idea that music performance should be "hard." Interestingly, there are examples of musicians specializing in playing automatic instruments. One of the more well-known is Rex Lawson, who works as a professional pianolist. His performance setup is based on a standalone pianola mechanism attached to a regular piano (Peress 2004). Lawson argues that even though the piano roll supplies the notes, the performer does the rest of the performance, including creating the dynamics and tempo. The fact that he draws large audiences in regular concert halls attests to people seeing him as a performer.

The discussion of automatic instruments becomes more exciting when investigating self-playing instruments in which there is no direct energy transfer between performer and instrument. By "direct," I here refer to in-time and real-time performance. Self-playing instruments are constructed with a system for building up potential energy. This potential energy can then be transformed into kinetic energy playing back the music. The performer would not need to do anything beyond starting the mechanism. Many children's toys are based on this principle. You wind up a spring, and the toy plays a melody. Technically speaking, there is little difference between a music box played directly via a crank and another based on a wound-up spring. One has a spring that can preserve energy, and the other does not. In terms of performance, however, there is a difference. In a music box that relies on a direct energy transfer, the performer continuously controls its speed. This makes it similar to the pianola in that it is possible to create an "expressive" performance through tempo adjustments during playback. In a music box that preserves energy, on the other hand, the music box plays by itself. How does that influence our perception of those using the devices?

I find different opinions about automatic instruments fascinating. In many ways, these discussions precede similar debates about electroacoustic instruments. After all, if a music box is considered an instrument, and the person using it is a musician, we should think similarly about other systems that can play music: LPs, cassette players, CDs, MP3 players, and apps on mobile devices. Thinking about these devices as musical instruments and their users as musicians may seem strange, but it is the logical consequence of the argument. Hence the user of such an automatic instrument is a performer even though the only active musicianship involved is adding energy to the system. That the performer is left with relatively little control over the final musical output is not reason enough to be disqualified as a musician. Instead, automatic instruments can be seen as bridging the gap between a performer and a perceiver. Here the musicking quadrant can be helpful to approach the analysis of such devices. Instead of thinking about performers and perceivers as categorically different, the quadrant opens for musicking activities between these roles, as sketched in figure 7.9a. When playing on a music box, the person can be both performer and perceiver simultaneously. Also, the development

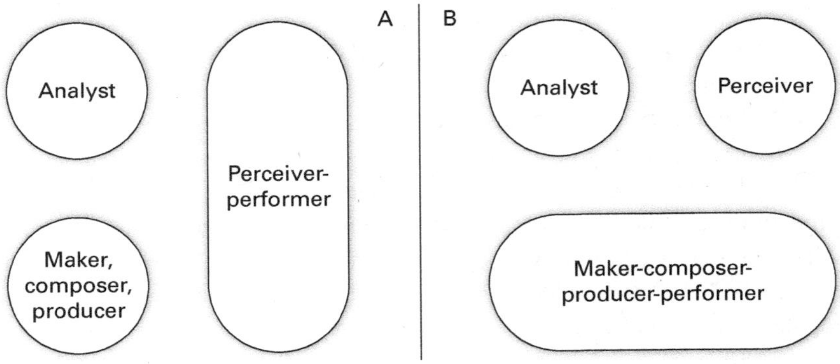

Figure 7.9
Automatic instruments challenge our thinking about traditional musical roles. The users of such instruments can be thought of as perceiver–performers (A) while the creators are a kind of "maker–composer–producer–performer" (B).

of such an instrument is blurred. Since the instrument is constructed with one or more musical pieces built into the physical design, one could also say that there is a blurring between the instrument maker, composer, producer, and performer, as sketched in figure 7.9b. We will see the same type of merging roles when considering various types of electro-acoustic instruments.

Conceptual Instruments

The final step in the continuum of action–sound separation in acoustic instruments may be somewhat puzzling. It is not about physical instruments at all, but about "conceptual" instruments. One such example is that of "imaginary" instruments. These do not exist in reality, only in our minds. Yet from a philosophical perspective, it makes sense to also include such instruments in this discussion. After all, everyone can imagine that they "listen" to an imaginary instrument. For example, think about a ten meter-long guitar with silver strings. Can you imagine how it sounds? Can you play a tune on it? Can you feel how it would be to touch the strings? While people's imagery of such an imaginary instrument would certainly differ, I believe that thinking about such instruments is also an embodied experience. The category of conceptual instruments would also include what could be called "impossible" instruments. The bike harmonium proposed by Carelman (1969) is one such example. It exists in drawings but cannot be physically produced.

Some Considerations

As the discussion above has shown, I support an inclusive approach to what should be considered a musical instrument. However, there are several holes in my theory. Let us consider some of them.

Orchestra as an Instrument

Can an orchestra be considered an instrument? I would say no, but before we completely abandon the idea, let us evaluate some variations of the question. By definition, an orchestra is composed of several instruments and several musicians. If there is only one musician with multiple instruments, this would usually not be called an orchestra. A borderline case would be a "one-man band," in which one person plays on numerous instruments. I would still argue that this is not an orchestra, since it is perceived as one person playing on multiple instruments. This also proves my argument that we perceive each of the instruments independently of each other.

Another case is that of a multiuser instrument, where you have one instrument with several performers. As argued above, I think that an acoustic church organ should be considered a multiuser instrument if one person pumps the air while another is playing on the keys. A more generally accepted example is probably a duo playing with four hands on the same piano. Then there are two people playing on the keys of the same instrument.

A more challenging question is whether a group of separate instruments, played by multiple musicians, could somehow be considered "one" instrument. This may seem absurd or irrelevant, but it is a good test case of my theory. The exact number of people and instruments does not matter for this argument, but for a common reference, let us think of a traditional symphony orchestra with a conductor. Suppose you listen to a symphonic concert on headphones, focusing on the sound of the music. In that case, there are good arguments for thinking about the orchestra as a combined sound-producing entity, thus one instrument. For this to work, however, the orchestra would have to work hard to blend. At best, this works effectively, but at worst, it falls apart, and the instruments are heard separately. It also depends on the music being played. Although several twentieth-century composers explored different spectral composition techniques, much orchestral music has been composed with a focus on exploring the interplay between various instruments (timbres) and instrument groups (textures). So even from a perceiver's point of view, I would argue that it is common to listen to the orchestra as a combination of individual instruments. This is particularly the case when watching the performance of an orchestra. Seeing the musicians play will make it very difficult to avoid separating the different auditory streams.

I doubt that many orchestra musicians think of the orchestra as one instrument. They have a close physical connection to their own instruments and continuously listen and engage with the others. On the other hand, a composer may think about the orchestra as a "vehicle for sonic expression." After all, the combination of different timbral qualities defines the orchestra's final sounding result. Orchestration is based on learning how to blend different instrumental timbres and build convergent or divergent sonic textures. Does that make the orchestra into one instrument for the composer? I would say no, based on the spatial and temporal disconnection between the activities of the composer and the orchestra. The composer typically works out of time related to the performance situation and is usually not involved in the performance. The composer will also have to think about the individual instruments when writing out the individual score parts.

The conductor is the last role we should consider when discussing the orchestra as an instrument. It is perhaps also the most interesting person in this context. One could say that the conductor "plays" the orchestra, at least in the meaning of controlling the final sonic output. This may, in fact, not be so far from the role of a pianolist for a pianola. Someone/something else is producing the musical output, but the conductor/pianolist brings the music "to life." At least, we may argue for an apparent temporal connection between the conductor and the orchestra. However, the conductor has a disembodied role when it comes to sound production. Even though the conductor controls the orchestra musicians, it is the musicians that produce the sound on their instruments.

All in all, I would argue that it is difficult to say that the orchestra is an instrument. However, it could be considered a "meta-instrument." It also makes sense to think of both the composer and the conductor as "meta-performers." They are indirectly involved in music production, with different levels of spatiotemporal distance. As such, their roles may be seen as similar to music producers and engineers. This is something we will get back to when looking at electro-acoustic musicking in the next chapter.

The Environment as an Instrument

As discussed previously, and as summarized in figure 7.2, the environment is also an essential element in the mental imagery of (musical) sounds. Humans are generally very capable of separating a sound source from the environment in which it is played. That is also why we can hear a violin playing in anything from a small room to a large cathedral. It has been argued that this is easier to do when the environment has natural characteristics. Traer and McDermott (2016) conducted a study where they recorded the impulse responses of several hundred rooms, after which they altered the frequency responses slightly. They found that people were better at discriminating sound sources from rooms when the reverberation had natural characteristics. One explanation for this may be an internalization of environmental acoustics through development or evolution.

The sonic properties of a room (the "acoustics") have in many cases a significant impact on the resultant sound. Both musicians and audiences know to appreciate a good-sounding venue. As Howard and Angus (2007, 307) phrase it:

> Reverberation time is an important aspect of sound behavior in a room. If the sound dies away very quickly we perceive the room as being 'dead' and we find that listening to, or producing, music within such a space unrewarding. On the other hand when the sound dies away very slowly we perceive the room as being 'live.' A live room is preferred to a dead room when it comes to listening to, or producing, live music.

Some rooms are built with particular acoustical features. The room acoustics of churches, for example, are of critical importance for the music being performed. Plainchant sounds utterly different inside and outside of reverberant spaces. This is also an example of how music is made for particular environments. For example, churches were a test bed for the development of poly-choral singing, "cori spezzati," exploring the spatial distribution of sound (Bryant 2018). However, not many people would argue that a church is an instrument, although it is an exciting line of thought.

Several musical instruments have been designed for particular purposes and venues. The development of louder instruments in the nineteenth century, for example, was driven by new and larger concert halls (DeNora 1995). One could also argue that the concert halls could be enlarged because of new and louder musical instruments. Still, most instruments are constructed independently of the performance venue. From a perceptual point of view, I would argue that even though space is essential for shaping the sound of many instruments, it is difficult to say that a room is part of our notion of the instrument. A notable exception is church organs that are custom built for a particular venue. Here the room could be argued to be the main resonator of the instrument.

The Philips pavilion at the Expo '58 in Brussels is a famous example of the merging of architecture, composition, and instrument design. Here, Iannis Xenakis and Le Corbusier carefully designed a space together with the composed music. As such, one can consider the building itself as a carefully crafted part of a large and complex instrument. The music was played through loudspeakers, though, so this would at best be an example of a hybrid instrument. There are also examples of natural and created sound sculptures based on acoustic sound making. We have a beautiful sound sculpture located at the west entrance of the Nationaltheatret train station in Oslo (figure 7.10). Here people passing by will experience a spectacular flutter echo, which invites for sonic exploration. I always pass by whenever I am around that part of the city. It is fascinating to see how both young and old make sounds and listen to the strange echo before smiling and hurrying down to their train. Such sound sculptures are fascinating, but are they instruments? Some could be, as they may contain both excitatory and modulatory parts. Then they could just be seen as vast acoustic instruments. Others, like the sound sculpture in Oslo, are primarily sound modifying. The sound production is done by people passing by. On the other hand, one may argue that exciting a string that resonates in

Figure 7.10
The west entrance of the Nationaltheatret train station in Oslo is constructed as a sound sculpture featuring a spectacular flutter echo.

a guitar body is no different from exciting air molecules that resonate in space. So we talk about borderline cases of how far we want to stretch the notion of an instrument.

From Sound Maker to Music Maker

An emerging topic is whether music makers should be considered instruments. I would say yes, but many people refuse my arguments. Some seem to think that instruments are sound makers that should be hard to play. The argument goes something like this: everyone can play a music box; hence it cannot be an instrument. But why not? What is wrong with anyone being able to play an instrument?

Music boxes are examples of instruments in which a complete musical piece is embedded in the instrument. It is possible to change between different tunes in some music boxes, but they can still only play what is preregistered in the "score." This is what Rowe (1993) would call a "score-driven" system. The performer can control the loudness and tempo, but not much else. Automatic instruments are music makers that result from the gradual mechanization of musical instruments throughout the centuries. While one may think of such automatic instruments as entirely different from other acoustic instruments, I see more of a continuum. For example, a self-playing piano is a pianola with a system for energy preservation. The pianola is a piano with a system added for recording and playing musical scores. A piano is a harp with an added keyboard mechanism. And a harp is a development of basic string instruments made from a stick and string.

All instruments reflect the (musical) culture they were developed within. Or, in the words of Magnusson (2018), "Instruments are impregnated with knowledge expressed as music theory. . . . They explain the world." The piano's keyboard layout has successfully established the twelve-tone equal temperament system as a core part of much of Western music. This is because of the piano's capabilities but perhaps also its lack of ability to explore other tunings and microtonal explorations. Fretless string instruments, on the other hand, afford much richer tonal explorations. Still, one may argue that the construction of string instruments with four to six strings and some common tuning standards also influences the music. It is, of course, possible to tune the violin differently, but, as De Souza (2017) argues, some internal musical logic is inevitably built into the construction of the instrument itself.

The gradual "musicification" of instruments from sound makers to music makers is also the story of an increasing separation between action and sound. Consider the difference between whistling and playing the silver flute. The first flutes allowed richer and louder sounds than what you could do solely with the mouth. Making holes in the flute allowed for playing discrete pitches. Adding keys to cover the holes ensured more precise playing of those pitches. More mechanical parts made it possible to expand the register. Most instruments have developed over the centuries to allow for more accurate control of intervals, specific timbres, and simultaneous tones (Keislar 2009).

We can also talk about a gradual "verticalization" of musical content as instruments have moved from sound makers to music makers. Most instruments with a low action–sound separation are based on playing melodies or rhythmic structures, what we could call "horizontal" musical elements. String-based instruments, such as pianos and guitars, make it possible to develop harmonic structures through chords. Some organs allow for playing additional tones through doublings, and in accordions, you can play complete chords with one finger. All such inventions have given musicians the ability to produce more "vertical" musical complexity in the form of more and more complex tones. All these developments continue in the world of electro-acoustic musicking, which is the topic of the next chapter.

8

Action–Sound Mappings

"We hear new sounds, we see familiar sounds coming out of unfamiliar devices, and unfamiliar sounds coming out of familiar interfaces," writes Kvifte (2008, 45) in a discussion of the changing nature of musical instruments. Indeed, today's new instruments are in many ways different from the traditional instruments we are accustomed to performing and perceiving. Yet, as the discussion in the last chapter showed, traditional instruments also differ. New acoustic instruments are still invented, but there are many more new electro-acoustic instruments. The latter is based on designing and constructing action–sound "mappings." In electro-acoustic instruments, the interaction is created by transferring electric signals (analog or digital) in a chain of interconnected devices, such as sketched in figure 8.1. I will argue that there are both practical and conceptual differences between instruments based on couplings and mappings—and electro-acoustic instruments allow for even more explorations into both sound making and music making.

From Electro to Acoustic

Before moving on, it may be worth repeating my rationale for using the term "electro-acoustic instrument." The "electro" part of the term refers to the need for electric power to produce sound in such instruments. The "acoustic" part refers to the fact that these instruments produce audible sound. I stress this because I find that the final sounding sound is often neglected in the literature on instruments. Even the legendary synthesizer builder Bob Moog "forgot" about the sound-producing element when he summarized the three determinants of musical instrument design (Moog 1988):

> The first is the sound generator; the second is the interface between the musician and the sound generator; the third is the . . . visual reality of the instrument.

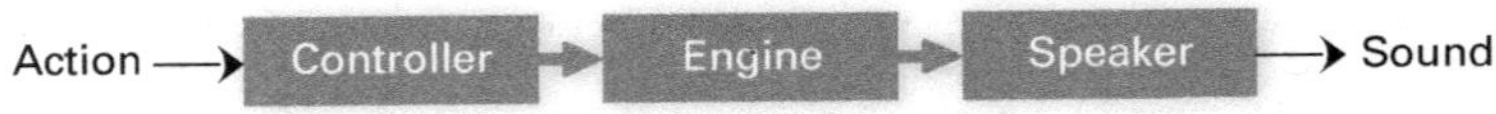

Figure 8.1
An electro-acoustic instrument can be broken into three core components: controller, sound engine, and speaker.

I agree with Moog that the sound generator—what I will call a "sound engine"—is a central part of an electro-acoustic instrument. As sketched in figure 8.1, the sound engine is controlled by a performer's actions. These actions are picked up by sensors in a physical "controller." Finally, the sound is produced by a "speaker" element, typically connected to an amplifier.

What differentiates electro-acoustic instruments from acoustic instruments is the lack of a direct, physical energy transfer from the performer to the final sound. The performer uses energy on the controller, but this is not the same energy used to produce sound coming out of the speaker. Instead, the instrument relies on energy from an external power source. This is not a problem per se, but it is a different concept—and technological construction—than what is found in an acoustic instrument. In electro-acoustic instruments, the parts in the chain are usually functionally independent. This allows for customization since controllers, sound engines, and speakers can be chained differently. While this is one of the strengths of electro-acoustic instruments, it is also one of their conceptual drawbacks.

Miranda and Wanderley (2006, 4) argue that mappings between the performer's action and the sound-generating device are central to digital electro-acoustic instrument design. I would argue that mapping is also vital for analog electro-acoustic instruments, although there are some differences between analog and digital circuitry. However, these differences are much more minor than the differences between electro-acoustic and acoustic instruments. Action–sound mappings are designed. One may say that the couplings found in acoustic instruments are also "designed" in that various sound-producing objects are put together in a particular way. In acoustic instruments, however, the resultant sound is based on the inherent physical properties of its sound-producing objects. In electro-acoustic instruments, the mappings created between a controller and a sound engine are arbitrary and may confuse both performers and perceivers. As the last decades of research into new interfaces for musical expression have shown, many instrument builders strive to create meaningful mappings in their electro-acoustic instruments.

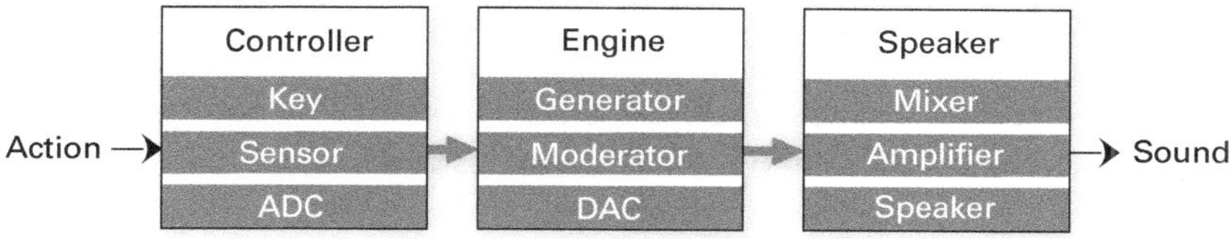

Figure 8.2
A schematic overview of some parts making up a controller, sound engine, and speaker.

The Parts of an Action–Sound Chain

Before moving on to some concrete examples, we will look more closely at each of the parts of the mapping chain described above. These are summarized in figure 8.2. We will consider the sound-producing parts first and then move on to the controller and mappings between action and sound.

The Sound Engine

The sound engine is at the core of any electro-acoustic instrument. After all, this is where sound generation occurs. Broadly speaking, we can talk about analog and digital electro-acoustic instruments. That is, instruments that produce sound through either analog circuitry or digital code. I have chosen this terminology to avoid the terms "electric" and "electronic" instruments. Any instrument powered from electricity could be called electric. An electronic system, on the other hand, is based on manipulating electrical current through switches. This is how most analog electro-acoustic instruments are constructed. Many of these are built around one or more oscillators, an electronic circuit that produces a periodic, oscillating electronic signal. An oscillator converts direct current from a power supply to an alternating current signal, resulting in a sine tone or square wave. There are only a few instruments that would be categorized as purely electric. One of these is the Victorian synthesizer (Collins 2009), which we will get back to soon. Most analog electro-acoustic instruments, however, could be classified as electronic.

When it comes to digital systems, it is worth mentioning that technically speaking, such a system would be a subcategory of both electric and electronic. Digital systems are electric since they run on electricity, and they are electronic because they are based on electronic circuitry. The difference is that the electronic circuits used in digital instruments operate with binary logic (ones and zeros). An electro-acoustic instrument based on digital processing is built around a discretization of sound signals. This involves an analog-to-digital converter (ADC) for sound recording and a digital-to-analog converter (DAC) for sound production. It could be argued that the discretization process in digital instruments makes them inferior to the (in theory) continuous signal of analog instruments. However, today's ADC/DAC converters run at sample rates and with bit depths higher than the limits of the human auditory system. So there is no practical difference between analog and digital circuitry any longer. The main theoretical difference is that the signal is passed through the system in analog circuitry without any discretization. This has some practical implications. Digital sound processing has revolutionized the industry, allowing for noiseless recording, mixing, copying, and transmission of musical sound. Interestingly, despite the many positive sides of digital sound processing, there is a continuous demand for analog devices. Some people argue that imperfections of analog circuitry make the sound "richer" (Jenkins 2007). Others complain about the delay—measured as the latency—of digital systems (Jota et al. 2013).

Acoustic instruments can often be broken down into sound-producing and sound-modifying elements. The same is the case for electro-acoustic instruments. Many synthesizers are built around the concept of sound "generators," such as different types of waveforms (sine, square, noise, and so on). The sound coming out of the generator can be fed into a "moderator." For example, a filter can shape a sound's timbral qualities, while envelopes and low-frequency oscillators shape its temporal development. Many systems also play with feedback loops between generators and moderators. This allows for creating complex timbres, textures, or rhythmic patterns.

The Speaker

In his foreword to the book *Material culture and electronic sound* (Weium and Boon 2013), Brian Eno eloquently describes his thoughts on the importance of speakers:

> So modern electronic instruments are incomplete without their "bodies"—the amplifiers and loudspeakers by which they're heard. ... Before the amp and speakers; it's all about numbers.

Eno writes "numbers" here, by which I take it that he refers to digital devices. However, also for non-number-based (analog) electro-acoustic instruments, there is no sound without a speaker. As sketched in figure 8.2, a speaker also typically needs one or more amplifiers to produce audible sound. I have also added a mixer to the chain, since that is often needed to route signals correctly. The mixer may also act as a sound-modifying device on its own.

Generally speaking, we may separate speakers that produce sound waves in the air from those that make sound through a physical object's vibrotactile stimulation. The first is what is found in most loudspeakers, in which the loudspeaker cone moves and creates sound waves in the air. When people ask me why I am interested in music-related body motion, I ask them to touch a speaker element. Feeling the vibration of a speaker element is a way of "touching" the sound. It is also an efficient reminder that sound is vibration. Normal loudspeaker elements produce vibrations in the air. Vibrotactile speakers, on the other hand, produce vibrations in physical objects. Here the final sound is heavily influenced by the acoustic properties of the object in question. For example, adding vibrotactile speakers to an acoustic guitar body is different from trying to excite a concrete wall. Most electro-acoustic instruments are still based on loudspeakers producing sound in the air. However, there has been an increased interest in "hybrid instruments" in recent years (Lähdeoja 2015; Eldridge and Kiefer 2017; Melbye and Ulfarsson 2020). These are based on creating vibrations in acoustic objects based on analog and digital circuitry connected to vibrotactile speakers.

Headphones is a speaker category that receives little research attention despite their musical impact. Many people perceive music through headphones, and many musicians playing electro-acoustic instruments spend a great deal of time rehearsing with headphones. Many musicians also perform with headphones these days. Then I include all sorts of headphones: in ear, over ear, around ear, and bone conductive. The latter is a particular category, but what the others have in common is that they primarily transmit sound directly to the ear. This is different from other types of speakers, which also transmit sound waves to the body. Everyone that has stood in front of a large PA system can attest to the bodily sensation of the sound waves. Given their widespread usage, I am puzzled why there are so few studies on the differences between headphones and speakers. For that reason, we carried out an experiment at the University of Oslo in which participants stood still while listening to music through either headphones or speakers. Interestingly, we found that people systematically moved more when listening to headphones than speakers (Zelechowska et al. 2020). This, we believe, can be explained by the "closing off" of one sense. Just as one gets more unstable by closing the eyes, putting on headphones removes much contextual information about the space one is standing in. This again may affect the balance. In our experiment, we used loud but comfortable sound playback levels from both loudspeakers and headphones. So the energy of the sound waves presented through loudspeakers was not enough to physically move people. It would be interesting to know more about how much people are actually moved when listening to loud music in front of a large PA system.

Bone-conducting headphones are an exciting in-between speaker technology. This is not a new type of device, but we have seen more commercial interest in such products in recent years. Particularly for people that need to keep attention to their audible surroundings, such as joggers, this type of device is useful. Bone-conducting headphones are not placed over the ears but rather on the chin. The vibrotactile stimulation makes it possible to hear sound and music quite satisfactory, although not with the same fidelity level as one would expect from a regular pair of headphones. I find it both strange and fascinating to wear such headphones. You have an intimate sound experience while also hearing everything else in the environment. Such headphones allow for perceiving the "outer" world while at the same time having an "inner" experience of music. While headphones are attractive in many ways and deserve more research attention in general, I will primarily focus on loudspeakers in the rest of this book.

Looking at the sketch of equipment in figure 8.2, the chain may be short if the sound engine is connected directly to the speaker. However, the sound may be passed from the sound engine to a mixer, connected to an amplifier and the speaker. In active speakers, the amplifier is built into the speaker cabinet, but it is still there. There may be multiple layers of mixers, DI boxes, and amplifiers going into a large speaker rig in larger concert setups. This usually also impacts the sound in one or more ways. If the chain is based on passing analog signals, this may "color" or distort the sound. Using digital mixers helps in avoiding such artifacts but may introduce latency. Also, like for acoustic instruments, it is essential to consider the space within which an instrument is played. Electro-acoustic instruments are also played in spaces with acoustic features. In addition, it is common to add virtual space information to the mix. We will look more at how such equipment chains may increase (or decrease) the perceived spatiotemporal distance in chapter 9.

The Controller

The controller is the device that captures the performer's action and passes it on to the sound engine. At the core of a controller is a sensor that converts physical energy into an electric signal, what Bongers (2000) calls the "sense organ" of the system. In some cases, the user may be in contact with the sensor directly. More often, the sensor is covered by some mechanical moving parts, such as the keys on a keyboard-based controller. As reviewed by Medeiros and Wanderley (2014), numerous sensor types are used for musical applications. They can be discrete, such as on/off-type capacitance sensors, or continuous, such as force-sensing resistors (FSRs). Light sensors and temperature sensors can be used, although they are often too slow for musical interaction. Then various infrared sensors may be more practical, either used as on/off switches or for continuous control. As described in chapter 4, various types of motion tracking systems are also increasingly used in interactive music systems.

No matter what type of sensor is used, it needs to be connected to the rest of the circuitry through a sensor interface. A sensor interface typically contains an analog-to-digital converter (ADC) that converts the electric sensor signal to a binary control signal in a digital instrument. For most physical controllers (for example, keyboards), there would usually be many mechanical parts, each connected to one or more sensors. There may also be multiple sensor interfaces on the inside, although the end user may only see a single stream of controller values that can be mapped further.

There has always been much experimentation with different types of controllers for both analog and digital instruments. Miranda and Wanderley (2006) suggest four categories: (1) augmented musical instruments, (2) instrument-like controllers, (3) instrument-inspired controllers, and (4) alternate controllers. This is a good summary of controllers found in the music technology research community. However, looking at commercially available controllers, the vast majority of products fall into one of these categories (or a combination of these):

Keyboard-like controllers: typically modeled after the piano, with its black and white keys

Mixer-like controllers: built using the "knobs-and-sliders" paradigm found in most mixers

Pad-like controllers: controllers built around a grid of touch-sensitive buttons

Acoustic instrument–like controllers: for example, drum kits and wind controllers

As described in chapter 6, the MIDI standard has been successful in allowing for setting up any type of connection between controllers and sound engines. But its focus on impulsive note-on/note-off messages may also be one of the reasons why most of the above-mentioned controller types (minus the few wind controllers out there) are built around sensors that work well within a MIDI-based control paradigm.

Fortunately, the music technology research community is actively developing new prototypes, with the annual International Conference on New Interfaces for Musical Expression (NIME) being the most central dissemination channel (Jensenius and Lyons 2017). We will look more at some of my contributions to this community in chapters 11 and 12.

Mappings

As mentioned above, we may think of mappings as the "glue" between the three main parts of electro-acoustic instruments: controllers, sound engines, and speakers. In its technical sense, mappings describe individual control parameters' connections to individual sound synthesis parameters. We may also think about mappings in a techno-cognitive sense: the experienced relationships between actions and sounds. This is the way I use the term when talking about action–sound mappings. The technical and conceptual levels are, of course, related. Ideally, a conceptual mapping idea is realized in the technical mapping.

Levitin et al. (2002, 171) start their discussion on strategies for developing new instruments with three postulates:

- Suitable mappings must be found between a musician's gesture and the control of various aspects of a musical tone (Cadoz et al. 1984; Wanderley 2001).
- Gestures are motions of the body that contain information (Kurtenbach and Hulteen 1990).
- Mappings are best when they are intuitive and when they afford the maximum degree of expression with minimal cognitive load (e.g., Keele 1973; Mulder et al. 1997).

Although I use slightly different terminology, these postulates resonate well with my action–sound theory. When it comes to creating mappings, this may happen at multiple stages throughout the design phase. At the most technical level, deciding which sensors to include in the controller may be seen as part of the mapping process. While any standard industrial sensor may be used for musical applications, Marshall and Wanderley (2006) have argued that it is essential to evaluate sensors with respect to their intended application. For example, a bend sensor may be used to control anything from pitch to loudness. Still, it may not be the best sensor considering the controller's actions or the parameters available in the sound model. An intuitive instrument design should be based on developing a controller that captures the intended actions. This should be followed by choosing a sound synthesis model with input parameters that fit the controller's outputs. The result would be action–sound mappings that work well for the performer and are intuitive for the perceivers.

On the sound engine side, there are many initial mapping decisions to be made. A typical sound synthesis model consists of many technical parameters. These do not necessarily correlate directly to perceptual qualities. A standard solution to master such a complex system is to create presets that define a set of parameters that work well together. These presets may then be used for further exploration of the sound synthesis model. However, the drawback of such a design decision is that users may end up using only presets. This may or may not be a problem, but it should be considered during the design phase.

While many commercial electro-acoustic instruments are heavily preset based, there has been much exploration with different mapping strategies in the music technology research community. When I teach electro-acoustic instrument design, students often start by creating one-to-one mappings between the controller and the sound engine. Connecting a slider to a sine tone generator is one of the most classic examples. Such a basic mapping makes sense from a technical perspective. One could even argue that the sensor and sound engine afford such a one-to-one mapping. For the user, it may make sense to have direct control of the sound, although it seldom leads to impressive musical results. Kvifte (1989) has argued that many acoustic instruments are, in fact, based on "coupled" mappings. For example, the breath of a clarinetist may control the timbre, loudness, and vibrato of the sound simultaneously. Similarly, these sound parameters may also be controlled by lip pressure. Many acoustic instruments are based on one-to-many or many-to-one mappings. In studies of people's experiences of different types of mappings, Hunt and Wanderley (2002) found that people often find such coupled mappings to be more intuitive and exciting than more straightforward one-to-one mappings. The study confirmed that what might seem to be an efficient mapping strategy from an engineering point of view may not be the best from a musical perspective.

There are also numerous examples of more complex many-to-many mapping solutions. For example, Momeni and Wessel (2003) suggested using a three-dimensional geometrical representation to control a highly multidimensional sound synthesis engine by interpolating values coming from a graphic tablet. Dahlstedt (2001) developed a system in which an evolutionary algorithm would generate new presets by keeping some information from the previous presets. In this way, it was possible to generate a set of presets with similar characteristics to the presets of the "parents." Another approach to controlling complex sound models was suggested by Bevilacqua and Muller (2005), who used statistical models to learn mappings between multidimensional control parameters and sound synthesis parameters. In that system, the user would make an action with a controller and select some sound synthesis parameters that the action should control. The model would also interpolate between different actions, which would result in combinations of the assigned sound synthesis parameters.

Although artificial intelligence (AI) methods have been around for decades, it is only more recently that it has been possible to work with machine learning in musical applications (Iñesta et al. 2018). Such models can be trained before a performance, but tools like Wekinator make it possible to use machine learning in real time (Fiebrink and Cook 2010). This allows for creating complex many-to-many mappings during performances. The development of musical deep learning methods will probably further drive the mapping possibilities in future instruments (Briot et al. 2019). A compelling aspect of such AI-based systems is that they do not require knowledge about the mapping parameters. The mappings are created automatically based on examples given by the user. That may work well, but it also provides the user with less direct control of the instrument.

The above-mentioned approaches are often based on "point-wise" mappings, defining points (presets) in a multidimensional space relating control and sound synthesis parameters. Van Nort (2010) argued that one should not only define what to map, but also how it is mapped. The trajectories between points in a mapping space may change the "feel" of a controller. Anyone who has switched between logarithmic and linear mapping knows how different they feel in practice. So one could argue that a controller's perceived expressivity is linked to the dynamic quality of its mappings.

"Physical modeling" is a conceptually different approach to mapping than the examples mentioned above. Creating models that reflect the underlying physical sound production may be seen as an "ecological" approach to sound synthesis. In such models, the control parameters are inspired by physical actions and objects rather than some abstract technical parameters. Physical modeling synthesis was first proposed by Karplus and Strong (1983), and their original model is commonly referred to as Karplus-Strong string synthesis. This physical model aims to simulate the pluck of a string by sending a burst of white noise into a filtered delay line. The result is a remarkably rich sound, particularly given the few elements of the algorithm. What is fascinating about physical modeling, as opposed to techniques such as additive synthesis, is that the user can "excite" the model by sending noise bursts into the delay line. This can be seen as analogous to performing multiple "plucks" on a physical string. Since the Karplus-Strong model is based on a delay line, one may continue to re-excite the model before the sound has decayed. This will build up the resonance in the model, just as you would experience with an actual resonating string. As such, a physical sound model behaves quite similarly to real-world sound excitation. This is quite different from, for example, sample-based sound engines, in which one always talks about the "polyphony" of the system as the main limitation for the sonic richness that it is possible to achieve.

Since a physical model is developed based on the imagined physical properties of the actions and objects involved, it calls for a direct mapping process in which the controller's physical outputs may be mapped directly to the input parameters of the sound model. An early example of a physical modeling approach to musical instrument design is the CORDIS system by Cadoz et al. (1984). While physical modeling has yet to reach its full commercial potential, research progress has been made with the development of "digital waveguide synthesis" by Smith (1992) and "physically inspired stochastic event modeling" (PHISM) and related controllers developed by Cook (2001). There are also examples of how physical modelling can be used creatively to set up unintuitive mappings. Wang et al. (2020) explored using a single-reed woodwind controller with a bowed string synthesis engine. This is an odd combination, but less so than using an impulsive controller with a sustained sound model. At least, the setup allowed for continuous control of the sound engine.

Much more could be said about mappings; in fact, Baalman (2022) has written a whole book on the topic. My main point is to stress that mappings are integral to the design of an electro-acoustic instrument. It is also a central part of the creative process when working with electronics in a musical context. While mapping has received more interest in the research domain in recent years, there is still much work that needs to be done both theoretically and practically. There is much creative potential in the development of new and better mappings between action and sound. There are also numerous scientific insights to be gained through studies of mappings. After all, mappings tap into some of the underlying mechanisms of our perception of actions and sounds in general.

Feedback

Feedback is another topic that has received relatively little attention in the design of *inter*active instruments. We may identify three layers of feedback in acoustic instruments: haptic, visual, and sonic (Kvifte and Jensenius 2006). The performer will receive some direct feedback from the instrument. For example, hitting the key on a piano will result in tactile and haptic feedback through the finger. One will also get visual information by seeing the key move down. Finally, one will hear the sonic result from the interaction. The sonic feedback may not only be from the vibrating string. Sometimes one may hear sounds from the piano mechanism itself. Other times one may hear the sound of hitting a piano key with a fingernail.

Similar types of multimodal feedback can be found in electro-acoustic instruments. Bongers (2000) argues that there is both active and passive feedback in electro-acoustic instruments. Here "active" can be seen as the feedback designed into the electronic part of a system. There is also passive feedback in the mechanical parts of the instrument. This may or may not be intended by the designer, but it is still there and needs to be considered when analyzing an instrument. I always think about this when I play on one of my old (digital) keyboards. It has much dust between the keys, which results in audible acoustic noise in the keyboard mechanism.

As sketched in figure 8.3, the audible sound is the central feedback modality in a musical instrument. A synthesizer sounds utterly different, whether connected to a pair of headphones, a pair of small office speakers, or a PA system. This sonic feedback is, obviously, essential for the performer. For instruments with built-in speakers, the user may also experience haptic feedback from the sound vibrations in the instrument body.

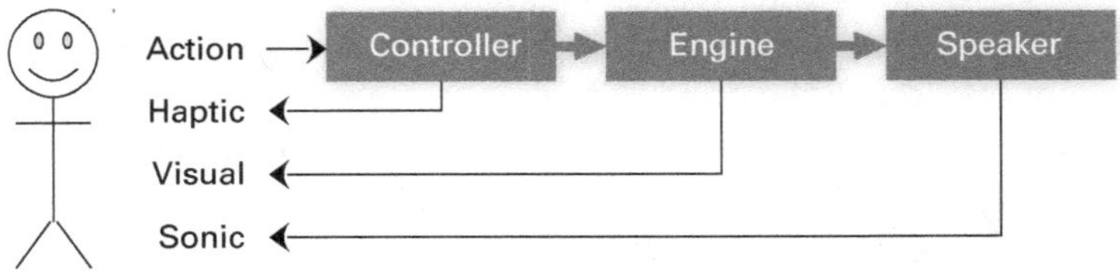

Figure 8.3
A sketch of three different types of feedback in an electro-acoustic instrument.

Visual feedback is another critical element of most electro-acoustic instruments, arguably more than in acoustic instruments. Such visual feedback can be in the form of light (for example, blinks or changing colors). Screens have also become prevalent in electro-acoustic instruments and are growing larger every day. In fact, in tablet-based and mobile phone–based instruments, the screen has taken over the entire interface. Barbosa (2019) argues that visual feedback needs to be direct, intuitive, and rapid. This helps compensate for the lack of physical feedback. The downside is that visual feedback makes it necessary to look at the instrument. This may potentially lead to communication challenges if the performer continuously stares at their instrument. Another problem with visual feedback is that it is often binary in nature. Many electro-acoustic instruments have lights that turn on and off to display whether some function is activated. This is a "cost-effective" feedback strategy that has the advantage of quickly giving the user control of many different types of parameters. The downside is that such feedback may tend to focus on displaying information about technical parameters. Having static binary feedback may also discourage continuous control.

The third feedback layer sketched in figure 8.3 is that of haptic feedback from the controller. Many electro-acoustic instruments have a mechanical interface, which may act back on the user. For example, the keyboard of an electro-acoustic instrument provides some tactile and haptic feedback in itself. Some high-end commercial instruments, like digital pianos, have active haptic feedback, and we will look at one such example in chapter 10. The interest in haptic interfaces has increased rapidly in the music technology research community (Papetti and Saitis 2018), so I expect more of this feedback modality in the years to come.

Braasch (2009) reflects on the complete neglect of olfaction in musical instrument designs. The sense of smell is much older than many other modalities from an evolutionary point of view but is rarely used in human-computer interaction. This is not only about digital scent technology; the smell of an instrument is primarily related to its physical materials. As for most other technologies, metal and plastic have dominated the construction of electro-acoustic instruments. There are some examples of the usage of other materials. Wood has seen a revival, and there are also examples of the use of leather in instruments (Favilla et al. 2005). This certainly helps in creating a richer olfactory (and tactile) experience.

While there may be exceptions, I would argue that an instrument's feedback level is directly linked to its experienced action–sound separation. Feedback is something you get for "free" in acoustic instruments, particularly those with a smaller separation between action and sound. On the other hand, each feedback layer needs to be designed and constructed in electro-acoustic instruments. The larger the action–sound separation of the instrument's construction, the more focus needs to be put on creating feedback. A good example here is the visual feedback design of the "buttons" of the Ocarina mobile phone–based instrument. Wang (2018) describes how carefully these were designed, with many visual layers, each having a different purpose. Such detailed designs are essential when the rest of the action–sound chain may be suboptimal. Then multimodal feedback may be used as a way of improving the overall experience of the instrument.

Action–Sound Separation

Since the construction of electro-acoustic instruments is fundamentally different from acoustic instruments, I do not think it is possible to use the same categories when talking about an instrument's action–sound separation. I will instead propose six categories that are more focused on the type of mapping process involved:

Embodied: The body is directly involved in the sound production or modification, such as in the theremin or cracklebox.

Analog: There are continuous (electric) relationships between action and sound, like in most analog synthesizers.

Digital: There are discrete (sampling-based) relationships between a physical input device and controller.

Virtual: There are multilayer relationships between action and sound, such as in apps and web-based instruments.

Automatic: The instrument can play on its own.

Conceptual: There is no physical instrument to interact with; the instrument and its sound production are only happening within your (embodied) mind.

These can be laid out schematically (figure 8.4), referring to whether there is a tighter or more loose action–sound separation. In the following, we will look more closely at each of these categories before discussing some more complex examples.

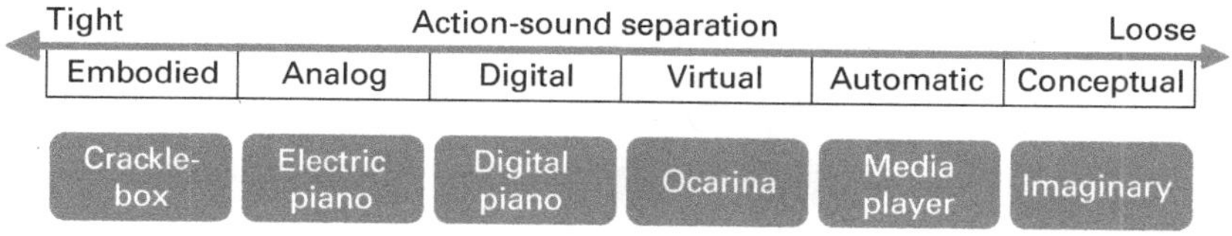

Figure 8.4
A sketch of different types of action–sound separation found in electro-acoustic instruments.

Embodied Instruments

Are there any electrically based embodied instruments? No, if we think about embodied acoustic instruments such as the voice. Here we will also leave out hybrid instruments based on amplifying body sounds in one way or another. Still, some electro-acoustic instruments are more embodied than others. The theremin is one of these, played by moving the hands within the range of two metal antennas. One of the antennas controls the frequency of the sound and the other the amplitude. It may seem like a contradiction to call the theremin an embodied instrument, while at the same time, it is also one of the earliest analog instruments and the first "touchless" instrument. However, the presence of the performer's body is necessary to both produce and modify the sound. When considering the whole chain from action to sound, I find it interesting that the original theremins were constructed with a built-in speaker. In modern-day versions of the instrument, the speaker has been removed.

Another instrument that could be considered partly embodied is the cracklebox (figure 8.5). This instrument was invented by Michel Waisvisz at STEIM in Amsterdam in the 1970s. Its construction is based on a single operational amplifier and a few transistors. The box has six metal contacts on top, and sound is produced when the performer touches the metal contacts to shortcut the circuit. The human body becomes a part of the circuit and determines the range of possible sounds. The sound comes from a small speaker in the box, so the cracklebox is a complete electro-acoustic instrument.

Waisvisz (2004) describes the invention of the cracklebox as inspired by touching the inside of his father's short-wave radio receivers. This resulted in alterations of the sound, through which he could start to play directly with the sound waves. From my action–sound perspective, I find it fascinating to read the reflections of Waisvisz (2004) on the physicality of the cracklebox:

> Finger pressure curves are very basic information standards. The act of applying physical effort through touch is empirically "known" to all human beings.

Figure 8.5
A cracklebox is played by touching the metal contacts on the top, thereby shortcutting the circuit to produce sound.

> The listener can feel the performer's touch and recognize the effort. The handling of physical effort is part of a universal language.

I find the cracklebox an inspiring instrument, mainly because it is unpredictable. When first touching it, you may not get any sound at all, particularly if you have dry fingers. It relies on some sweat or moisture to be conductive. The next challenge is to make consistent sonic results. I often find myself sitting with it for long periods, exploring new types of "sonic shapes." The timbral quality of the sounds is usually not that interesting; however, the close relationship between finger action and sound results in human-like sonic shapes. I also like that the cracklebox has a built-in speaker. This creates an immediate connection to the sound being produced, further strengthened by the zero latency of the simple analog circuit. The use of wood and metal in the construction additionally helps create the feeling that you are playing a "real" instrument, not just a toy. It may be the combination of all these factors—the tactility, performability, and playfulness—that together make for what I will later describe as an "affective" experience.

Analog Instruments

Analog electro-acoustic instruments come in all sorts of shapes, sizes, and levels of complexity. I mentioned the theremin and cracklebox as

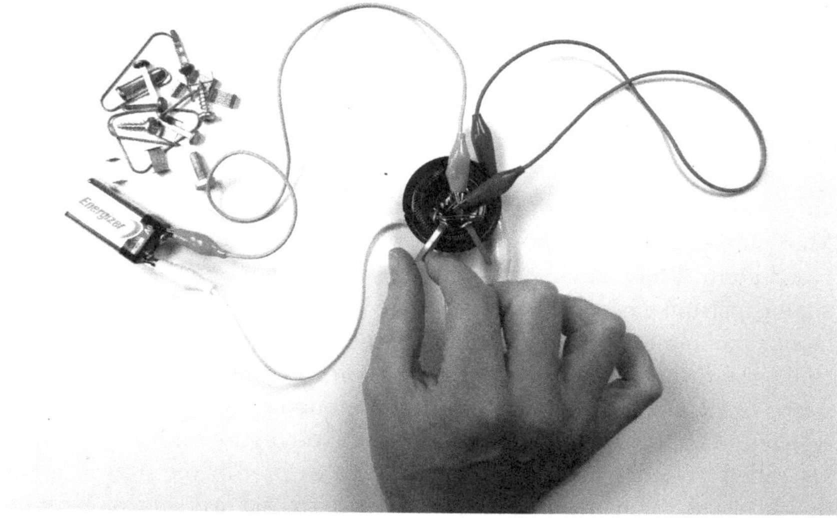

Figure 8.6
A Victorian synthesizer is possibly the most simplistic electric instrument. It only contains a speaker element, a battery, and cables between them. The instrument can be played by adding metal pieces to the circuit.

partly embodied instruments, but they are certainly examples of analog instruments. Another of my favorite analog instruments is, in fact, even simpler: the Victorian synthesizer (Collins 2009). This is one of a few genuinely electric instruments since there is no electronic circuit involved. It contains so few parts that everyone can build it themselves. It only consists of a battery, a speaker element, and cables connecting them, as shown in figure 8.6. Sound is produced when the circuit is closed between the battery and the speaker element. This creates a single "spike" of sound and immediately invites further sonic exploration. There is no latency in the chain, and you can feel the vibration in the speaker element when it happens, so there is both sonic and haptic feedback provided to the user. It is possible to develop small self-playing music machines by adding more speakers, cables, and extra metal pieces into the chain (paper clips work great, for example). I find the Victorian synthesizer to be excellent in teaching principles of sound technology. It demonstrates the core elements of an electro-acoustic instrument, and literally, everyone can make it work. With some practice, it is even possible to come up with relatively complex musical results. I once had a group of students that managed to create a sophisticated rhythmic structure by connecting the various parts I had given them in class.

There are numerous examples of more complex analog instruments; see, for example, the excellent overview of analog synthesizers by Jenkins (2007). When browsing catalogs of commercially available analog instruments, one realizes that the keyboard quickly made its appearance. Famous examples of early analog instruments—such as the Ondes Martenot—rely on a piano-like control surface. As we know, keyboard-based control has remained remarkably popular in later generations of analog instruments and has also been the de facto standard in digital interfaces. Of commercially successful instruments with keyboard control, one could mention the Hammond organ, Wurlitzer piano, and Moog synthesizers. I find it interesting that while the Hammond and Wurlitzer are often referred to as "organ" and "piano," respectively, many Moog instruments are called "synthesizers." This may be seen as part of the shift of attention from instruments being sound makers to music makers. Many of the early synthesizers could be thought of as something in between: "sound modifiers."

It is also worthwhile reflecting on the use of the term "keyboard." The word itself refers to the controller of an instrument, as you would find in a piano or organ. Synthesizers, however, come in all shapes and do not necessarily need to have a keyboard-like controller. So what is the difference between playing a "synthesizer" and playing a "keyboard"? I have heard people argue that the main difference is that synthesizers do not have speakers built in, while keyboards do. The same people may also disregard such keyboards as toys for kids. In my thinking, the two terms are incomparable. One describes the controller (keyboard), while the other the sound engine (synthesizer). Numerous instruments with a synthesizer as a sound engine are not controlled with a keyboard interface. There are also examples of keyboard-based instruments that are not based on sound synthesis, such as sample-based devices. Nowadays, it is also common to buy keyboard controllers with no sound engine or speakers.

The gradual modularization of electro-acoustic instruments—both analog and digital—has changed how music is made. The first analog devices were standalone instruments, complete with a control interface, sound engine, and built-in speaker. Then the speakers were "removed," before the controller was also split from the sound engine. Finally, the sound engines were split into many small components. This gradual modularization was at least partly driven by the ability to route control voltage signals in different ways (Bjørn and Meyer 2018). With an increased standardization among manufacturers, it became possible to send electric control messages between devices. The development of digital MIDI signaling reduced the need for passing electric signals between devices. Analog instruments saw a general decline after the introduction of digital devices in the 1980s. Still, analog devices kept being developed and used, and we have recently witnessed a revival (Barlindhaug 2019). Several manufacturers have taken up production of old analog models or developed affordable versions of old synthesizer concepts. There is also a growing interest in modular, analog synthesizers made accessible by the standardization into Eurorack modules. Also, the "circuit bending" communities have embraced using and modifying analog electronics for musical applications (Skjulstad 2016). So analog electro-acoustic instruments are here to stay.

As we are getting to the end of this section on analog instruments, some readers may wonder why I left the electric guitar out of this discussion. That is because I am thinking of the electric guitar as a hybrid instrument, based on the amplification of an acoustic instrument. We will return to a discussion of the electric guitar and other hybrid instruments in chapter 9.

Digital Instruments

Digital electro-acoustic instruments differ from analog in that they are based on a discretized signal chain. This is done by using ADCs to convert control signals into a digital representation and the use of DACs to convert digital signals into audible sound. In my taxonomy, I separate "digital" and "virtual" instruments. This may seem like an arbitrary division, but the idea is to consider devices that have some kind of mechanical controller (such as a physical keyboard) separately from devices played through an intermediary control layer (such as a touch screen).

Digital instruments were gradually introduced to the market in the 1980s, with the Yamaha DX7 being the first commercially successful digital synthesizer (Shepard 2013). The possibilities of digital instruments have increased rapidly. What has remained remarkably constant, however, are the controllers. Even though we see some new explorations into multi-touch controllers, most commercial devices still rely on keyboards and variations of the "knobs-and-sliders" paradigm. The conservatism of manufacturers may (at least partly) be explained by the limitations of the MIDI standard. As discussed in chapter 6, the note-on/note-off-type messaging is designed for keyboard-based instruments, and the control message paradigm lends itself well for control by buttons and knobs. When there is a whole technological ecosystem built around this standard—which works well in many ways—it is challenging to think anew.

One of the benefits of MIDI is that it allows for modularizing setups. This modularization began with analog instruments and has become more prevalent in the world of digital instruments. On the positive side, one can use multiple sound engines with one controller or numerous controllers with one sound engine. However, this flexibility often leads to more complexity and difficulties in getting things to work. I have spent countless hours reading manuals and fiddling with settings to pass signals from a controller to a sound engine. Setting up for a sound check is also much more time consuming with modular setups than when using self-contained instruments. So the modularization has a downside when it comes to the level of "plug and playability" of instruments. It also leads to an increased action–sound separation, mainly when using different sound engines with the same controller.

In the taxonomy, I do not separate hardware-based and software-based digital instruments. One could argue that all digital devices are hardware based since they have a computer inside. However, there are differences between instruments based on microcontrollers versus general-purpose PCs. For example, instruments built around microcontrollers can often be more efficient for the task at hand despite their inferior technological specifications. DSP-optimized chips and their single-purpose design can compensate for the lack of computing power in such embedded devices. That said, PC-based digital instruments have become so robust and powerful that they are currently used in many contexts.

Let us consider the setup of a typical "laptop musician." This usually consists of a MIDI controller connected to a laptop running one or more music applications. The laptop is also connected to a sound card, passing the sound to a mixer and speaker system. Many things have improved in the world of computer music over the years, but I am unsure whether I spend more or less time troubleshooting my software-based instrument setups than my hardware-based ones. Computers usually give better error messages than hardware devices, but more things need to be connected to make the system work properly. Whenever I switch to a new laptop, I realize how many drivers and software packages need to be installed to get the different controllers and sound cards to work correctly. Constant software updates help fix bugs and add new features, but also lead to new problems and additional troubleshooting. Sometimes I am lucky and get things up and running quickly. However, I still think we are far from having true "plug and playability" in the world of laptop-based musicianship.

In addition to the "synthesizer-type" and "laptop-type" digital devices mentioned above, we should not forget about the digital instruments that aim to mimic acoustic instruments. In terms of sales numbers, I would guess that this is the largest category by far, of which digital pianos are probably the main driver. Several manufacturers have put in much effort to create instruments that are as "acoustic-like" as possible. This may be seen as an example of instruments that focus on decreasing the experienced action–sound separation and is something we will get back to in chapter 10.

Virtual Instruments

Virtual instruments are also based on digital signal processing, so they could have been included in the category of digital instruments. Still, I think they differ from an action–sound point of view. A piano app on a mobile phone is an example of what I would call a virtual instrument. More precisely, it is not the app itself that is the instrument, but the app inside a physical phone with a touch screen for input and a speaker for output. As such, a phone with a piano app is a complete instrument according to my definition. What makes such an app-based instrument different from a MIDI controller connected to a laptop is that there is one more abstract "layer" in the interaction: the app is played by touching "keys" on the screen. Thus the screen acts as a mediator between the finger and the virtual keyboard on the screen. Therefore, this interaction has one more degree of action–sound separation than a digital instrument in which you press on a physical key.

The same is true for web-based instruments, which you typically play by clicking with your mouse on "keys" in a browser window. Here the mouse is the physical interface, which is further mapped to clicking on the virtual key on the screen. Web-based instruments often also add an extra layer of abstraction on the sound-processing side. The Web Audio API has made it possible to develop powerful browser-based instruments (Smus 2013). Many of these rely on sound processing on the local machine. There are also examples of instruments in which all the sound processing happens remotely. In some cases, even remote hardware devices can be controlled, such as in the network control of Joe Paradiso's modular analog synthesizer (Mayton et al. 2012). Both technologically and conceptually, this leads to a large action–sound separation. Such setups also create more latency than if all the processing would happen locally. The exception is whether the sound processing is so heavy that it could not run in real time on a local computer. Then there may be benefits of "outsourcing" the processing to a remote server that can perform in real time, though slightly out of time.

While there are certainly some challenges regarding the increasing action–sound separation in virtual instruments, there are also several benefits. Server-based instruments may be more accessible to many people. There is no need for particular hardware on the user side; hence, users can start making sound without having to set up and configure anything. Large servers may also do much heavier processing than what could be done on a regular PC; thus, the final output sound could be more complex. However, server-based instruments allow for engaging in entirely new forms of musicking. Wang (2018) writes about the popularity of Ocarina, one of the first iPhone instruments. The sound making part of Ocarina is clever. The microphone is used to pick up the user's breath and is used for sound production. The sound modification is done by touching virtual buttons on the screen. However, the most exciting part of the instrument is its ability to connect to a global community of other Ocarina players. This brings in a new social dimension and allows for large-scale collaborative musicking.

The Ocarina is an excellent example of a virtual instrument that embraces the possibilities afforded by new technologies. Many other virtual instruments are more focused on recreating the past. Just in the same way that we have seen a lot of digital instruments that imitate acoustic instruments, we now see a lot of virtual instruments that mimic analog and digital instruments. There are virtual versions of old analog synthesizers, with the complete look and functionality of the hardware device. Instead of patching physical cables or touching knobs, the user controls the devices with a mouse or a touch screen. I find this somewhat peculiar. It is interesting from historical and educational perspectives, but I am more interested in developing and using new paradigms with new technologies than in recreating the past.

Another example of an innovative virtual instrument is Bloom by Brian Eno. This is also an iPhone-based app in which the user creates sound by tapping on the screen. Eno and Chilvers (2008) describe the instrument as "an endless music machine, a music box for the 21st century. You can play it, or you can watch it play itself."

The two modes of Bloom allow for different levels of musicking. When started in "playback mode," it can be considered a self-playing instrument, similar to a music box. The "performance mode" allows for direct

Figure 8.7
The Oslo Mobile Orchestra performing a piece using the performance mode of Brian Eno's Bloom (bottom). Each performer used a mobile phone connected to a handheld speaker (top). *Source*: Ståle A. Skogstad.

control of the tones being played. Here the user can touch the screen to control the pitch of the tones. So there is a clear and immediate control of individual tones. The underlying sequencer will repeat the same tones after a few seconds, and the user can add new tones to the mix. The user can switch between different settings, in which the scales and colors will change, but they are all laid out so that the instrument will always sound "nice." I still remember my then one-year-old daughter's fascination as she pressed the screen and listened to the sounds. It both gave a sense of being in control (performing) and experiencing (perceiving). As such, Bloom shows how the traditional roles of musicking are in play. This was also Eno's intention, describing it as creating "gardening" experiences (Clark Estes 2018):

> Imagine if composing could be more like gardening than architecture.... You do control the input, but you don't control the output.

While Bloom was primarily developed for individual usage, we enjoyed using it in performances with the Oslo Mobile Orchestra. As shown in figure 8.7, this ensemble was based on performing in the tradition of a "marching band," similar to other mobile phone orchestras (Oh et al. 2010). Each performer had a mobile phone, which was connected to a handheld speaker. This made it possible to freely move around in the room while keeping a sound source close to the performer and giving the audience a spatial experience of the sound. Each mobile phone speaker was not particularly powerful, but having many of them spread around a space resulted in a submerged auditory experience. When it came to controlling the instrument, Bloom is so easy to play that I could teach the concept to new ensemble members in minutes. One person that performed with us in a concert once objected that this was not a "real" performance. She argued that only practicing for five minutes before going on stage was an example of "fake" musicianship. Yes, playing a concert with Bloom is different from playing a classical violin quartet piece. The exciting thing, however, was that audiences were thrilled about our mobile phone concerts. I have never received more performance requests than with that ensemble. So what we think of as "true" performance, and what people find musically interesting, may not always be the same.

Automatic Instruments

Only a few acoustic instruments can play on their own, but most electroacoustic instruments can. For example, most digital keyboards have a "demo mode" where you can listen to different tunes stored in the instrument. I still have fond memories of the demo songs on my first keyboard from the 1980s. To my parents' despair, I used to play these songs over and over as a child. That particular instrument is now at our family cabin, and my daughters are the ones to keep pushing the "demo mode" button while dancing to the tunes.

While automatic playback is possible on many electro-acoustic instruments, it is primarily an add-on feature to demonstrate their capabilities. Their initial design is still meant for being played by humans. Other instruments are made primarily for automatic performance. The playback mode of Bloom, for example, was intentionally designed as such. But also Bloom's performance mode could be considered a combination of virtual and automatic instruments, or perhaps more as an "active listening" device. What is becoming tricky from a classification perspective, then, is to what extent music recording and playback devices should be considered instruments or not.

As described in the previous chapter, there are numerous examples of automatic acoustic instruments. These instruments were mechanical at first but then eventually became electrified. Without covering the whole story of recorded music—there are many excellent resources on that, such as those by Wurtzler (2007), Katz (2010), and Devine (2019) to name just a few—the twentieth century saw an incredible development from the first player pianos to gramophones with wax rolls followed by the LP, cassette, CD, and MP3 players and leading up to today's variety of streaming-based devices. These music storage devices are based on different technologies, but they share the same function of allowing for playing back recorded music.

In a review of musical agency in human–computer interaction, Tatar and Pasquier (2019) propose a taxonomy from "purely reactive" agents on one side to "completely autonomous" on the other. The autonomous instruments are arguably the most extreme case of automatic instruments. A former colleague of mine, Risto Holopainen (2012, i), defined these as "computer programmes that generate music algorithmically and without real-time interaction, from the waveform level up to the large scale form." He has continued to explore such instruments in both his research and composition practice. In the most extreme case, the computer can be programmed to create an instrument on its own, compose a piece for the instrument, play it back, listen to its creation, evaluate it, and use its aesthetic logic to reiterate and develop new instruments and music. While this may seem a bit far fetched, we will probably see more examples of autonomous instruments playing what Harper (2011) calls "infinite music."

I often hear people talk about autonomous instruments as synonymous with instruments driven by AI. An autonomous instrument could be based on advanced machine learning, but it could also be based on rule-based algorithms. If one argues that the latter should be considered AI, then autonomous instruments could be considered AI-based instruments. However, there are also cases where AI is used in musical instruments without them being autonomous. There may be machine learning algorithms on the sensing side, clustering algorithms in the sound engine, and evolutionary algorithms in the melody generator. Still, the instrument could be controlled by a human. See, for example, all the different approaches presented in Miranda (2021) to get an overview of the current state of AI-based musicianship.

Autonomous instruments may seem like the ultimate example of a complete action–sound separation. However, using such instruments could also be an efficient way of generating musical ideas that could serve as a creative starting point for human musicians. Drum machines and accompaniment systems have been available for decades. Today's autonomous instruments are building on such systems and pushing them some steps further. Some people are afraid that AI-based instruments will take over the jobs of composers and musicians. What is clear is that AI has already entered most parts of the music world and will continue to develop in many directions. I see this more as an opportunity than a threat. Someone needs to develop the underlying algorithms and put them to use. This requires new types of musical knowledge and experience. Instrument makers, composers, producers, and performers are an important part of the transition to tomorrow's musicking.

Conceptual Instruments

Some of the most extreme cases of autonomous instruments could perhaps also be considered conceptual instruments. Imaginary electro-acoustic instruments can give rise to vivid musical imagery. In some cases, they can even be built. Lepri and McPherson (2019) describe an experiment in which people were asked to create fictional, non-sound-producing instruments. Mockups of these instruments were later used to study how the designs could tell about the musical background of the inventor.

One example of an instrument that is both real and imaginary at the same time is *Volume 3: Everything You Love Will One Day Be Taken From You* by Yann Seznec (2019). This is the third volume of a set of three instruments called *The Book of Knowledge of Impractical Musical Devices*, inspired by the thirteenth-century *Book of Knowledge of Ingenious Mechanical Devices* by Ibn al-Razzaz al-Jazari. Seznec set forth to create three instruments shaped like books and that, in various ways, play with our understanding of time, space, and sound. In the third book, he explores the concept of the degradation of sound. It is inspired by Alvin Lucier's *I'm sitting in a room* from 1969. In the piece, Lucier recorded his voice and played it back multiple times in a reverberant space (Hasse 2012). Each playback and recording led to changes in the sound, based on the strengthening of the resonant frequencies in the room. At the end of the piece, one can only barely recognize the rhythm of the original speech; all the frequency content has been disturbed. Seznec's instrument is playing with the same idea in that the user only has one button to press. That button is playing back a sample, but at the same time, it is also replacing the same sample with a slightly distorted version of itself. So by listening to the sound, one will also gradually destroy it. The instrument exists as a digital instrument, although few people get to play it in person. For me, and most others, it is equally important as a conceptual instrument.

Some Considerations

As the above discussion showed, creating a taxonomy that covers all sorts of electro-acoustic instruments is challenging. I have considered many different types, but there are also many open holes in my theory. Let us consider some borderline cases.

The Need for a Speaker

While my instrument definition is flexible in many ways, I am strict about the need for instruments to make sound. Therefore, it is necessary to consider the whole chain from action to sound when analyzing an instrument. That means that a sound engine, such as a synthesizer module, is not an instrument on its own. It needs a controller for capturing actions and transmitting control signals. It also requires a speaker to produce sound. Similarly, a MIDI controller is not an instrument on its own; it necessitates a sound engine and speaker. Furthermore, a speaker is not an instrument; it needs a controller and sound engine.

This way of thinking about an instrument may seem radical. For example, it means that all the "digital musical instruments" that rely on external speakers are not instruments on their own according to my definition. However, they can be considered as part of an instrument if we also count in the speaker. Then, it is possible to view the whole chain from action to sound, including cables and other necessary components to produce audible sound. If there is no speaker, there is no sound. If there is no sound, it cannot be heard by the performer or the perceiver. This is the same as having a violin without a bow and strings. The violin needs strings and a bow to produce sound and hence be considered an instrument.

Some people ask me why I am arguing so strongly for considering the whole chain from action to sound in instruments. They say that a DX7 will sound like a DX7 even if you connect it to different speaker systems. I agree that the DX7—and many other signature synthesizers—have unique sonic characteristics. Still, the final-sounding sound depends on the speaker system used. Playing through headphones is not the same as playing through a hi-fi system at home or a PA system in a club. The playback method largely influences the resultant sound. I still remember when I connected my first digital piano to a large PA system. Until then, I had mainly played it at home with headphones. Using it on a stage with sound coming out of a PA system was an entirely different experience. In short, it felt like a new instrument.

Being strict about considering the whole chain from action to sound may feel limiting. However, it is also a way of acknowledging the complexities of today's and tomorrow's musicking. It is easy to define the instrument when one plays an acoustic guitar. The same is not the case with a laptop-based rig or a telematic performance setup. Many computer-based setups rely on a range of connected hardware and software components. If we want to understand how the music is performed and perceived, it is necessary to consider the whole chain from action to sound.

Music Playback Devices

Let us return to the question of how a music playback device—whether in the form of an LP player or a media player on a mobile phone—could be considered an instrument. In chapter 7, we discussed how a pianola or music box could be thought of as musical instruments, even though they are based on playing back prerecorded musical material. Still, they are devices made for performing music in real time, albeit with relatively few control parameters available for the user. Modern-day music playback devices are similar but allow for more control of the final music. They can start and stop the playback, move back and forth in a song, and control the loudness through an amplifier. In some cases, they may have rudimentary tone control through "treble" and "bass" buttons. Sometimes they may even have a built-in equalizer. All in all, these control possibilities allow the user to shape the final output considerably.

There are only slight differences between an old-school gramophone player and a modern LP player. One relies on mechanical power from a wound-up spring, while the other runs on electric power. They both rely on discs with engraved musical information and a pickup that transforms the signal into audible sound. In my thinking, both gramophone and LP players can be thought of as musical instruments. In fact, LP players have been on stages for decades, in the hands of DJs. A typical DJ setup consists of two LP players, a mixer, a microphone, possibly some sound effects, a pair of headphones, and a PA system. Therefore, one may argue that for the DJ, it is the complete set of devices that make up the instrument. I also find it interesting that there are two sound-generating devices in a DJ setup: the headphones used by the performer to precheck the tracks and the PA system from which the final mix is played.

DJs make a living from playing (with) prerecorded music. Some create a playlist before a show, and the only thing they do in real time is to start and stop the playback. One could argue that most of the musical choices and work of such a "playlist DJ" are done out of time and possibly in non–real time with respect to the "now" of the performance. This resembles the role of a composer but could also be seen as somewhat similar to the way a conductor selects repertoire and controls the start/stop of the performance and sound levels. Other DJs are more active performers, selecting new tracks, carrying out sophisticated beat matching, adding vocals, and performing elaborate "scratching" techniques. Hansen (2010) has shown how scratching has developed into a skillful performance practice with a complex musical output.

The most experimental DJs, who are often referred to as "turntablists," typically work with all sorts of mediums under their turntable stylus (Smith 2013; Holmes 2016). Here the LP player can be considered a particular type of "mic-and-amp" setup since the stylus effectively works as a contact microphone. Many people would probably agree that a turntablist is, indeed, a musician performing with an instrument. However, the instrument in question—the LP player with the sound system—is the same. As such, it is hard not to classify this as an instrument. It was, after all, even designed and made for making music. The same can be said about the new digital music playback systems that DJs commonly bring on stage. Just as for LP players, they also allow for various types of real-time music making.

Many would probably agree that a DJ or a turntablist can be considered a musician. However, it is more radical to claim that everyone using such technologies is a musician. Are you a musician when you turn on music on your mobile phone? According to my definition, yes. A mobile phone with a pair of headphones can be considered a complete musical instrument. And if you start and stop the playback of music and adjust the volume, you are actively taking part in the music making. You have limited degrees of freedom available for controlling the sound, but you are still in control of what is going on. You can skip songs, change the volume, and modify the sound settings. This is different from standing on a stage in front of many people, but the principle is the same: you "music." Whether you are a performer or perceiver is not about what you do but about the role you take on. The new technologies continue to blur these lines.

The Laptop as an Instrument

Many electro-acoustic instruments are based on a computer in one way or another, whether it is a small embedded device in a digital piano, a laptop in a PC-based setup, or a high-end server in an online instrument. Kvifte (1989) reflects on the change of "computers" from being humans doing manual calculations to machines taking over the same work. In any case, today's machine-based computers run programs created by humans. So humans are still in control of the final output. In his discussion of "machine musicianship," Rowe (1993) discusses the differences between score-driven and performance-driven systems. As mentioned in the previous chapter, a music box, pianola, or LP performance can be thought of as score driven. On the other hand, a performance-driven system relies on the performer to make musical decisions.

Nowadays, laptops are ubiquitous for music making and are, in many cases, part of an instrument. However, can a laptop be an instrument on its own? Fiebrink et al. (2007) argue that laptops have several built-in controllers (keyboard and mouse pad) and sensors (microphone, camera, tilt sensors, and light sensors). These can be used to control software-based sound engines. Laptops also have built-in speakers so that they can produce sound on their own. Laptops can even be considered mobile instruments since they can be moved around during performance. The same is the case with mobile phones and tablets, both of which can be considered complete musical instruments with relevant software installed. All of these devices—laptops, phones, tablets—fulfill my action–sound criterion: the ability to sense action and output audible sound. The same is not the case for a desktop PC, which needs to be connected to controllers and speakers to sense action and produce sound.

In my department, we have a growing number of students who reply with "laptop" when they are asked about what instrument they play. When someone tells me that they play "laptop," I always ask about what type of setup they use. Then they usually come up with a long list of different types of controllers, sound cards, mixers, and so on. When I ask them whether all of these devices should be considered part of their instrument, they agree that everything should be included. So it turns out that the term "laptop" is just a short form of a more extensive setup. This is similar to how someone playing percussion would not list up all the different sound-producing devices they may bring on stage. However, few laptop musicians include a speaker in the description of what they play. Some do, usually those that play guitar together with their laptop. Electric guitarists often think about a particular guitar amp as part of their instrument. They tend to be concerned with how the guitar sound blends with the laptop sound.

What people use a laptop for in performance varies considerably. Brown (2012) suggests thinking of three types of usage of computers in music: as a tool, as a medium, and as an instrument. Sound recording and processing are examples of how a computer is used as a tool. Listening to music streams would be a case of the computer as a medium. A typical laptop performance scenario would be considered an instrument. In my experience, these three categories tend to fuse in performance. As discussed above, a DJ may use the laptop primarily for playing back recorded music. A guitarist may use the laptop for playing some preprogrammed drum beats to improvise over. In other cases, a laptop may be used as a "looper" or to add effects. Finally, some may play the laptop as a "normal" instrument: real-time sound production and sound modification.

Laptop Orchestras

The development of the laptop as a musical instrument also sparked off an interest in laptop orchestras. While many had been using laptops in various performance constellations, the Princeton Laptop Orchestra (PLOrk) is often considered a forerunner in form and organization (Trueman 2007). The conceptual idea of modern-day laptop orchestras has been to explore the use of computer-based instruments in "classic" ensemble-based musicianship: trios, quartets, and even larger ensembles with tens of performers (Knotts and Collins 2014). These orchestras are often related to educational activities. They have the same hardware setup, including laptops, controllers, and speakers. I find it particularly interesting that so much focus is put on the speaker. This is perhaps the most distinguishing element of such laptop orchestras from other types of computer-based performance. The reason for this is that, just like in an ensemble based on acoustic instruments, the musicians need to hear their own sound and the other musicians' sound. This was—and to a large extent, still is, I think—a radically different way of thinking about the sound projection from computers. Rather than having all the sound come out of a large PA, a laptop orchestra's sound emanates from each musician's location. In the context of PLOrk and its sibling ensembles, this sound source would not only be in mono or stereo but projected from multichannel hemispherical speakers. As Smallwood (2017) argues, the ability to have local sound projection has turned out to be essential for the many different laptop orchestras developed over the years.

As laptop orchestras developed, so did the need for conductors and scores. In some cases, orchestras have worked with human conductors with experience from acoustic-based ensembles. However, due to the peculiarities of laptop-based instruments, conductors would often need to develop specific techniques for laptop orchestras. Using laptops on stage has also expanded the role of co-conductorship among the musicians. Instead of having a centralized conductor role, musicians can use messaging systems to conduct each other. This can also be combined with, or wholly overtaken by, computer-based conductors. The experimentation with different types of conductors has developed hand in hand with new kinds of scores. Since laptop-based music is rarely note based, it has been necessary to experiment with alternate scores. From an action–sound perspective, I find it interesting that many of these scores focus on performance actions ("move hand in a circle") or audible sound ("play bright sound"). In any case, they aim to provide an efficient representation of what will be played, either through representations on paper or dynamically presented on screens.

There had been computer-based ensembles also before the development of PLOrk-style laptop orchestras. However, the institutionalization of the new wave of ensembles helped establish the laptop as a viable instrument in many universities and conservatories. With complete and relatively easy-to-use individual setups, the structured approach to laptop performance has served as an entry point to computer-based performance for many music students. In many cases, such ensembles have also attracted nonmusic students, effectively acting as an interdisciplinary melting pot. This attests to the usefulness of computers in both traditional and nontraditional musicking. The development of repertoire, performance technique, and audience expectations has also helped shape the understanding that laptop-based instruments have come to stay.

One of the challenges I have seen with laptop-based instruments—particularly within the context of laptop ensembles—is how to handle changes in action–sound mappings from piece to piece. The physical setup stays the same in most cases, but it is common to change the mappings between pieces. This may not necessarily be a problem. Dramatic changes of action–sound mappings can be used creatively. However, the new mappings need to be carefully introduced to the audience for such changes to work effectively. It also helps if the mappings are based on a logical relationship between action and sound. Several laptop orchestras use different types of "gestural controllers" in performance (Tsoukalas et al. 2018). This adds a visual element to the performance. I also see that the development of new performance strategies within laptop orchestras has inspired the "gestural" vocabulary also for other types of laptop performance.

Live Coding

Next to laptop orchestras, there is another trend that has appeared in computer-based performance over the last decades, namely that of live coding (Collins et al. 2003). This is a particular type of algorithmic music (Dean and McLean 2018) in which the musician starts the performance with a blank page on the screen. Performances are based on writing code in real time, often projected on a screen for everyone to follow. This makes it possible for the audience to see the code development and how it turns into sound. In a way, this relationship between visible code and audible sound could be seen as the action–sound mapping in such a setup. A relevant question, however, is whether live coding can be considered an instrument? If so, I think it should be a virtual one since there are usually several layers of abstraction between action and sound. I would instead believe that it is the computer connected to a sound system that is the instrument. The code is part of the instrument but not an instrument on its own.

I am always fascinated by live coding performances. In many ways, it is a performance style that allows for seeing the performer's way of thinking about the generation of sound and musical structure before hearing the sound. As such, it may be thought of as what Chadabe (1997) refers to as the real-time equivalent of algorithmic composition. However, the question about how such a performance relates to the "now" is tricky. Sometimes there is an almost immediate sonic effect of the code being written. This could be seen as an in-time performance. Other times it can take seconds (or even minutes) before the written code materializes in sound. Such out-of-time performance can be an efficient way of creating expectations. The use of countdown messages, for example, is particularly fascinating because it shapes the performance. A statement such as "kill all in sixty seconds" creates an expectation of what will happen. Even though many in the audience may not understand everything happening in the code, such clear messages can be understood by everyone and help structure the performance. This is why many people who usually would not care much about electro-acoustic musicking can find live coding exciting.

Live coding is yet another example of how the traditional roles of musicking are in play. A live coder can be seen as a real-time composer, or an in-time improviser, creating musical sounds and structures on the fly. The performance has a real-time component, but there is often a delay from written code to what appears as sound from the speakers. It differs from many other types of music performance, which have closer connections between action and sound. This is not unique to live coders. Musicians using a drum machine or a DJ preparing a new track also perform in "almost real-time" mode. What is special about live coders is that they partly build their instruments while performing. They rely on some programmatic framework, but I would still argue that this performance style can be seen as a type of real-time instrument making. The performer develops the instrument on the fly. This is in line with what Green (2011) refers to as the "agility" and "playfulness" of musicking technologies.

Coming back to my initial question about whether a computer could be considered an instrument, my answer would be both "yes" and "no." Computers are general-purpose tools and can be part of an instrument in various ways. Laptops, tablets, and mobile phones can be self-contained musical instruments since they contain both controllers and speakers in addition to (software-based) sound engines. In most cases, however, I agree with Wang (2018), who suggests that computers should be thought of as "meta-instruments." Just in the same way that an orchestra can be used to create different rhythms, timbres, textures, harmonies, and melodies, a computer can also facilitate all of these and more.

The Studio as an Instrument

Can a studio be viewed as an instrument? Many would probably say "no," while others, including Bell (2018), have argued for the studio's instrumentality. There are several reasons why I find this to be a challenging question. It is not only theoretical either. In my department, we have had several discussions over the years about whether students can pass the instrumental entrance exam based on studio practice. When students argue for having the studio as their "main instrument," this is not only a question of identity. It is also based on the fact that they spend most of their time making music in a studio, and they would like to get individual teaching (and credits) for this studio work. On one side, this is a question about the institutional policy, which is primarily of interest for those involved. However, the question also raises a more conceptual discussion.

Without going too much into details, a studio—whether for recording, mixing, mastering, or all of these—is primarily meant for creating recorded music. In the beginning of the music recording industry, all of these were done in real time during the performance (Burgess 2014). As such, music recording was a real-time activity happening alongside the performance. Real-time music recording still occurs but is usually only the beginning of a studio process. The exception is when music is streamed live. In parallel to such real-time recording of performances—with or without much extra work in studios afterward, we have seen the development of a new studio-based music production practice: the "art of record production" (Frith and Zagorski-Thomas 2012). This studio-based practice requires its own set of technical skills, logistics of handling the different people involved, and the aesthetics of making the final music "product." The fact that one talks about a product, not a performance, is also why I believe that the art of record production is primarily a non-real-time activity. As such, one can argue that music production has more in common with composition than performance. The difference between the two is that a composer relies on a performer to play music at a concert. A producer includes the performer in the process. There are certainly performance elements in music production, although in different ways than when playing traditional instruments.

Sterne (2003) argues that musicians and engineers have bridged the gap between a musical instrument and their reproduction through recording. My intention is not to make an unnecessary divide between playing an instrument and working in the studio. Still, I think it is helpful to differentiate between real-time and non-real-time processes. There is a difference between performing on stage in front of an audience and performing in a studio. One is happening in the "now," and if you make a mistake, there is no way you can correct it. On the other hand, performing in a studio, you are used to making multiple takes, and the recording is only part of the process of making the final product. One could think about that as a performance on its own, but this performance is usually made only for the few, not for an external audience. I would, therefore, argue that "traditional" studio work more closely resembles composition than performance.

Things quickly become more complicated when considering a particular type of "studio": the digital audio workstation (DAW). Just in the same way that hardware-based instruments can reappear in a digital format, nowadays, a whole studio can manifest itself in a laptop-based setup. However, can a laptop-based DAW be considered an instrument? We could argue that the process of making music with a DAW is also happening in non–real time. Hence the DAW is not any more instrument-like than a large-scale studio. What makes DAWs different is that they have started to move on stage. Some of them also include several real-time features in the form of virtual instruments and sound effects.

If a performer plays on stage with a laptop (running a DAW), some controllers, and a sound system, it ticks all the boxes of what would be called an instrument. What makes it challenging is that non–real-time features may be used in real time. Confusing? Yes. Consider, for example, sampling as a technique. This is based on recording sound and then playing it back, either just once or in a loop. This is an example of how in-time and out-of-time performance meet in real time. Similar to live coders, such a performance would also be based on working with compositional strategies in real time. Starting up a drum machine or a looper or any other type of time-based sound engine would be an example of performing in the "future." A traditional musician is focused on performing action–sound couplings or mappings in the "now." A producer–musician would create sounds to happen in a "little while from now" or "sixteen bars from now." The possibilities are endless, which is both positive and negative. Magnusson (2010, 62) describes this as "the practically infinite expressive scope of the environment, sometimes resulting in a creative paralysis or in the frequent symptom of a musician-turned-engineer."

Numerous more examples could have been mentioned. The main point is that current music technologies, in which DAWs are an inspiring example, challenge a traditional understanding of musicking. The conclusion, then, should the studio be considered an instrument or not? My answer would be both "yes" and "no." It all depends on the context. In general, however, I would probably think about the studio as a kind of meta-instrument. Working in non–real time in the studio more resembles compositional practice. However, when a DAW is used in a laptop-based setup on stage, it can be thought of as part of an instrument. Looking back at the musicking quadrant, we see that the roles continue to blur.

9

Spatiotemporality

"Can you hear me?" This must be one of the most well-used phrases among people testing communication technologies. Sitting next to someone in a room, we never need confirmation that they hear us. But when communicating across cables, with few feedback modalities available, asking for confirmation is the best option. Feedback also concerns the separation between actions and sounds. The interaction feels immediate if the action–sound separation is low. As we have seen in previous chapters, new technologies have allowed for creating instruments with larger action–sound separation. This separation is partly caused by a longer latency between when an action is performed and a sound appears. The modularization of instruments has also increased the "distance" between action and sound in various instruments. New technologies challenge the idea of something happening "here" and "now." This is mainly the case when we look at telematic musicking. In this chapter, I will introduce the term "spatiotemporal distance" to explain the spatial dislocation and temporal lag between action and sound. The term can be seen as a further specification of how we can think about the separation of action and sound.

Sound Amplification

In his book *The Audible Past*, Sterne (2003) traces the history of contemporary sound reproduction to pre-electric times. He argues that the invention of the telephone and gramophone did not come out of the blue. Instead, these devices were the results of centuries of scientific discoveries and cultural developments. Sound reproduction technologies can be thought of as an extension of the human auditory system's "tympanic" principle. As Sterne (2003, 34) argues, "To speak of a set of sound-reproduction technologies as *tympanic* is to understand them

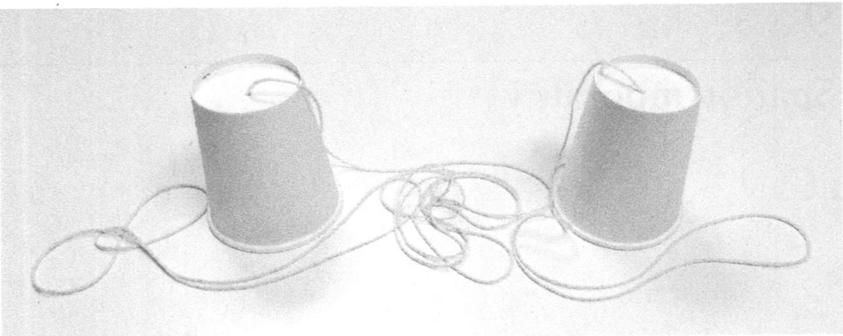

Figure 9.1
A paper cup phone is an example of an acoustic microphone and speaker.

as all functionally related, as sharing a set of common operational and philosophical principles, and, most important, as embodiments and intensifications of tendencies that were already existent elsewhere in culture."

This line of thought resonates well with my thinking about relationships between humans and technologies. It also aligns with the reflections made by van Eck (2017) in the book *Between air and electricity*. Here she argues that microphones—what she elegantly describes as "softhearers"—and loudspeakers also have a pre-electric history. Think about the "paper cup phone" that many children play with (figure 9.1). This is, in fact, an example of an acoustic microphone and speaker. The excitation of the "membrane" through talking on one side travels through the line and sets the paper cup on the other side in motion.

The paper cup phone is just one example of analog amplification. The megaphone is another—a device built for amplifying the human voice. One could also argue that most acoustic instruments have built-in amplification through their resonating bodies. The tube's size and shape in brass instruments are made to create as loud a sound as possible. Similarly, string instruments' body size and shape were refined for centuries to create the right sound level and timbral qualities. So most acoustic instruments do, in fact, already have built-in amplification.

The invention—or, perhaps better, evolution—of microphones and speakers can be seen as a disruptive music technology. Here I use "disruptive" to emphasize that these technologies radically changed how music was produced and performed. I would not go as far as van Eck (2017, 145) in arguing for microphones and loudspeakers being instruments, although she acknowledges that "they never manage to behave entirely

Figure 9.2
A typical chain in an amplification system, consisting of a microphone connected to a speaker.

like conventional instruments." Instead, I think about microphones and speakers as sound-modifying devices.

If we look more schematically at a modern-day amplification system, we find a microphone–speaker pair at its heart (figure 9.2). Such a microphone–speaker system is based on a sound–sound coupling or mapping. I think it makes sense to continue to differentiate between couplings and mappings. In the paper cup phone, there is a physical translation of energy from one cup to another, through the string. As such, the paper cup phone can be said to consist of a sound–sound coupling. However, most modern-day amplification systems require electricity to operate. The digital ones also rely on discretizing the signal through an ADC and a conversion back to sound through a DAC. That is why I prefer to think of these as sound–sound mappings. What differentiates a sound–sound coupling/mapping from an action–sound coupling/mapping is that they transmit sound from one side to another. One could argue that sound is, physically speaking, motion and could be considered an action. However, given the different frequencies of bodily actions and audible sound, I will leave that argument aside.

Figure 9.3 shows a sketch of a complete action–sound chain involving the amplification of an instrument. The instrument is played with actions, and its sound is fed into a microphone–speaker system to produce the final sound. One crucial element here is that the instrument also makes sound independently of the microphone–speaker system. This is indicated by the dotted line in the figure, showing that the instrument's initial (electro-) acoustic sound may also be audible. For example, if you play an acoustic guitar with a microphone in front of it in a small room, you hear a combination of the acoustic sound from the instrument mixed with the sound coming from the speaker system. On the other hand, nobody will probably hear the instrument's acoustic sound if you play an electric guitar on a big stage with a large PA system. Still, the acoustic sound of the guitar is there and is the source of the amplified sound.

As mentioned previously, electro-acoustic instruments may also make some purely acoustic sound. Such sounds typically result from the mechanical construction of the instrument and are not thought of as

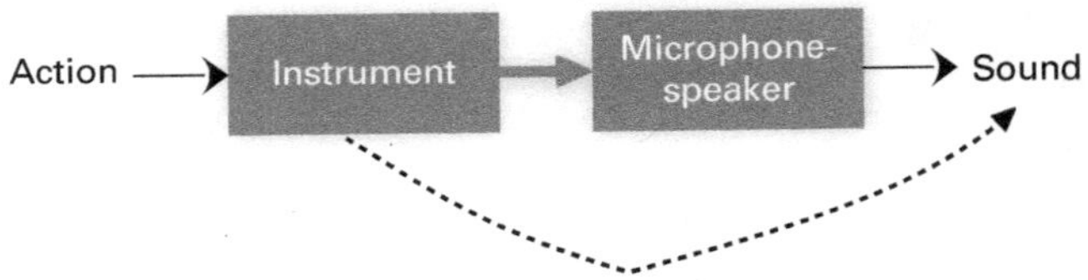

Figure 9.3
The sound from an amplified instrument consists of a mix of the sound from the microphone–speaker system and the direct (electro-)acoustic sound.

"musical." I always find it interesting to listen to people playing on a keyboard synthesizer with headphones. Hearing their finger tapping on the keys, the wobbling sound of the keyboard stand, and the squeaky sounds of the chair reminds me about the acoustic nature of "digital" musicking. There are even a few examples of how such unintended acoustic sounds can be used in instrument designs. For example, Dahlstedt (2017) used the acoustic sounds of a digital keyboard to excite a wave-guide string model in a hybrid instrument design.

Microphones and speakers were originally designed to make a louder sound, but they also "shape" the sound in various ways. The nonregular behavior of amplifiers and speakers may cause artifacts in the amplification process. For example, many guitarists care about their amplifier because it "colors" the guitar sound. The shaping of sound often also involves various types of sound processing devices. Figure 9.4 shows a more sophisticated sound–sound-signal flow, with a windshield attached to the microphone and a preamplification step. The processing stage can be analog or digital. If it is digital, it will rely on analog-to-digital conversion. Then, one or more filters or effects can be applied to the chain, such as equalizing, delay, and reverb before the signal is turned analog again through a digital-to-analog converter. Finally, we may think of the speaker part as a combination of mixer(s), amplifier(s), and the actual speaker elements.

It is relevant to reflect briefly on how one perceives amplified sound. In the terminology of Clarke (2005), we may say that hearing an amplified voice is part of determining what the sound is the sound *of*. We can distinguish between (what we believe is) the original sound source and its processing and amplification results. Sometimes the effects used are aimed at being as unnoticeable as possible, what Brøvig-Hanssen (2018) calls a "transparent" mediation. Other times, they are supposed to be audible. For example, using a vocoder is meant to be heard; it adds another layer

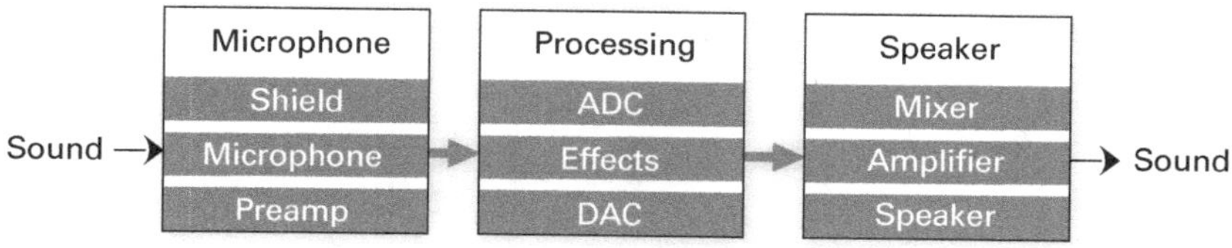

Figure 9.4
A possible amplification chain using digital effects between a microphone and speaker.

to the sound. This is an example of an "opaque" mediation. The idea here is that the perceiver focuses on the sound source in transparent mediation, while the mediation itself (the processing) takes on a meaning-bearing role in opaque mediation.

From a multimodal perspective, vision is also crucial for the way we hear. I recently went to see a theater play, which contained many musical elements. There were no musicians on stage, so I thought the music was prerecorded and played back in parts throughout the show. However, the "liveness" of the sound, including precise timing with the actors and some small playing mistakes, made me wonder if the music was actually played live. Even more strange was that I struggled to understand the instrumentation. I partly heard some acoustic instruments but also a lot of complex, digital textures. It was first at the end of the show I understood what was going on. Then a curtain in the back of the stage went up, revealing a small band. They performed on a combination of acoustic and electro-acoustic instruments and with lots of sound effects. Without seeing them play, it was challenging to understand what was going on. However, as soon as I saw the performers and their setup, it all made more sense. I adjusted my perception of the entire performance when I knew that real musicians were playing throughout.

Dislocation of Action and Sound

Microphone–speaker systems allow for increasing the physical distance between action and sound (figure 9.5). It is not uncommon to have a concert setup where a performer on stage talks into a microphone connected to a mixer in the back of the room many meters away. The sound engineer adds EQ, compression, and reverb in the mixer and sends the sound to the speakers. If the signal chain is analog, the distance traveled will define the signal delay. The distance matters less in a digital signal chain since the signal speed in Ethernet cables is much faster than in audio cables. However,

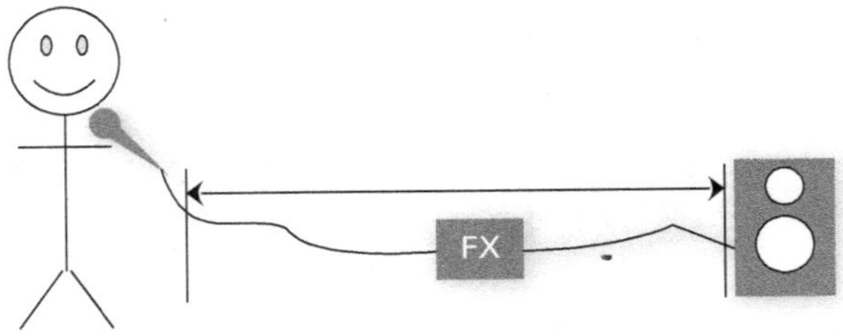

Figure 9.5
New technologies allow for an increasing physical dislocation of action and sound.

a digital signal relies on analog-to-digital and digital-to-analog converters, which may add extra latency.

The signal chain can be seen as the technical dislocation of action and sound. This technical dislocation may or may not relate to the dislocation perceived by the performer and perceiver. Even when there are long signal chains, a performer may have onstage monitoring in the form of a loudspeaker close to the microphone or an in-ear headphone. Thus, the perceived action–sound "distance" may be very short. The sound may also come out of speakers that are further away from the performer. It is not unusual to have speakers placed ten meters (or more) away from the performer on a big stage. Then the performers and perceivers will experience a larger spatiotemporal distance. I find that going to large stadium concerts is always a mixed experience. It is not uncommon to have multiple speaker sets spread out. This is good for providing better sound to people far from the stage, but it comes at the cost of hearing a mix of speakers with slight timing differences. In addition, there is a difference between the performers' actions on stage and a delayed version of the same actions projected on large screens. The result is a combination of multiple action–sound dislocations.

Let us consider a setup for a typical electric guitarist. This can be seen as a variation of the singer-focused microphone–speaker setup presented in figure 9.5. A guitarist is often standing in front of a guitar amp when performing. This serves the purpose of local monitoring but is also vital in shaping the sound. It is possible to connect the guitar directly to the mixer, in which one can also add effects to shape the sound. Still, many guitarists prefer to use a physical amplifier on stage with a microphone to capture its sound. Then the sound is amplified twice, first in the onstage amplifier and later in the PA system, such as sketched in figure 9.6. This leads to a relatively short spatiotemporal distance between the guitarist and the onstage amplifier and a longer one to the mixer and PA system.

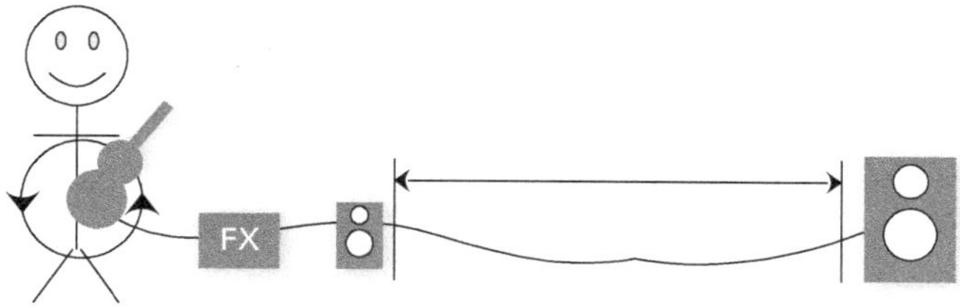

Figure 9.6
Telematic performances connect multiple physical locations.

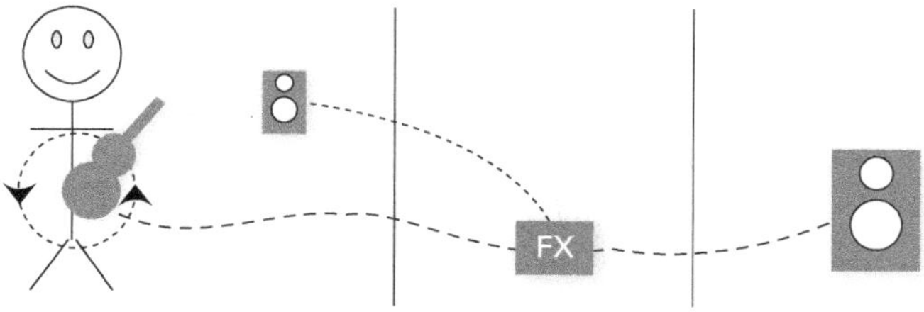

Figure 9.7
Server-based processing can be used in telematic performance setups.

With better and faster music technologies available, all sorts of routing possibilities emerge. Figure 9.7 is a sketch of an entirely dislocated setup, with which one could send unprocessed sound from a guitar to a cloud-based effects engine that processes the sound and sends it back for local monitoring and amplification. There are endless variations here. However, it is clear that monitoring—whether it happens through stage speakers facing the performer or through in-ear headphones—is becoming increasingly important when the use of amplification increases. A short spatiotemporal distance between action and sound is essential for the performer and arguably matters to the perceivers. If there is too much delay between what they see and what they hear, they will be confused at best but, at worst, lose interest in the performance altogether.

Latency

Latency is a much-discussed topic among music technology researchers. The technical latency can be measured as a signal's delay as it passes through the system. Perceptual latency—or the experienced delay—is related to the time it takes for you to do something and then experience the result. It is important to remember that latency is also an issue in acoustic musicking. For example, there is a considerable lag between pressing a key to hear the sound in church organs. Orchestra musicians are also used to dealing with latencies on stage. There is usually a considerable acoustic latency because of the time it takes for sound to travel from one side of the stage to the other. The speed of sound in air is around 343 m/s, dependent on altitude and temperature. So it takes the sound around 2.9 milliseconds to travel one meter. This means that on a ten meter-long stage, the acoustic latency is almost thirty milliseconds, just above the audible threshold. Although they are often negligible in practice, there are also action–sound latencies "within" acoustic instruments. In a detailed study of piano action by Askenfelt and Jansson (1990), they found that playing softly on a piano key may lead to latencies of up to thirty milliseconds just in the piano action itself.

How much latency can one tolerate? There is no clear answer to this question. In a study of musicians playing along to a metronome, Dahl and Bresin (2001) showed that musicians were able to compensate for latencies up to around fifty-five milliseconds; however, the sound's frequency content matters. In the music technology research community, ten milliseconds has been commonly accepted as a target threshold for a "computer's audible reaction to gesture" (Wessel and Wright 2001, 13). When designing new systems, such a threshold can be seen as a good "rule of thumb" but should not be seen as the ultimate answer to what can be considered an acceptable latency. Sounds with low-frequency content have long wavelengths and will usually have a higher experienced latency threshold. Therefore, it matters whether the instrument in question is used to play rapid high-frequency content or low-frequency drones.

Many electro-acoustic instruments—particularly computer-based setups with external controllers and sound cards—have high technical latencies and experienced delays. There may be more than a ten-millisecond delay in every step, adding up to hundreds of milliseconds in total. No wonder that many music technologists focus on minimizing the latency in their setups. However, the latency may not always be a problem. The latency may be less critical for performances based on repetitive processes (drum machines, samplers, loopers) or temporally long elements (drones, pads). Other times—particularly in telematic performances—one needs to accept the latency and see if it is possible to use it as part of the creative practice.

Telematic Performance

We have seen increased interest in telematic performance in recent years. This is sometimes called "distributed music performance," but I agree with Braasch (2009, 423–424) that "distributed" may be a misleading term. Any performance involving multiple musicians would be distributed in space. The critical point of telematic performance is that the performers (and perceivers) are located in physically separate locations. One could say that they are "colocated" in remote locations. Even though telematic performance has increased rapidly in recent years, it is, in fact, not a new way of performing. Holmes (2012) reflect on how telecommunication has been used for musical applications since the beginning. For example, one of the first synthesizers, the telharmonium, was performed over telephone wires.

I have been involved in telematic performances for nearly two decades and have seen the field expand in different directions. At first, we only passed control messages (such as MIDI or OSC) over the network, triggering or controlling sound engines on the other side. Then it became possible to pass audio in real time before video also became a reality. Since 2017, we have run the master's program Music, Communication and Technology at the University of Oslo. The program is built around the MCT Portal (figure 9.8), which can connect to other spaces around the world. Having a permanent telematic setup allows for continued exploration. In the past, we had to set up equipment, make a performance, and rig down again. Now we can explore telematics daily.

New technologies have allowed for increasing the distance between action and sound, both technologically and conceptually. Networking software such as Jacktrip (Cáceres and Chafe 2010) and LOLA (Drioli et al. 2013) have made it possible to explore high-quality and low-latency musicking. Now we also see the advent of digital mixers and commercial communication protocols such as Dante (Williams et al. 2011) and Network Device Interface (NDI) (Ellerbrock et al. 2004). These standards make it possible to send audio and video over networks with a high level of synchronization between streams. An interesting by-product of the development of network-based signaling is that it can also be used within a concert space. The result is a setup that is both more complex and simple at the same time. More complex, digital technologies (for example, a digital stage box and mixer) can replace analog technologies (such as cabling and DI boxes). This will probably be seen as a simplification from the sound technician's point of view. It may be a transparent change of technologies

Figure 9.8
The MCT Portal, a laboratory for low-latency audiovisual communication at the University of Oslo. Here set up for a performance with musicians in Oslo, Berlin, and Stockholm in 2021.

for the musician on stage, unless the added latency is perceptually challenging. It would probably also lead to a better overall experience for the audience. The change to network-based technologies may also lead to new audience experiences, such as what we have seen with the growth of online performances.

All in all, switching to digital technologies will, in many cases, lead to a better-sounding result and new musicking possibilities. However, there may be some added challenges related to an increased spatiotemporal distance. After all, musical coperformance relies on continuous *inter*action with the fellow performers. A temporal lag in the others' sound and image makes it challenging to perform together. Anyone who has tried to perform using a conventional video conferencing system has experienced that latencies of a few hundred milliseconds make it challenging to play normal rhythms and melodies together. This is particularly the case if there is also an additional variation (jitter) of the latency. Delays in the signal may work for regular conversations but not for time-critical music performance. Thus it is essential to reduce the latency in all parts of the action–sound chain.

Paradoxically, sometimes the experienced latency may go down in telematic performance. For orchestra musicians spaced out on a large stage (thirty meters wide), it takes approximately ninety milliseconds for the sound to travel from one side to the other. Suppose instead the sound from the instrument was captured with a microphone, digitized, sent through a network, and played through headphones to another musician. In that case, they might have a smaller perceived latency than through analog transmission. This is not to say that symphony orchestras should switch to perform telematically, but it brings another perspective to the latency discussion.

Latency may be considered a drawback, but it may also be used creatively. The piece/instrument Superstring Theories by Øyvind Brandtsegg and Bernt Isak Wærstad can be considered a "telematic instrument" (Brandtsegg 2022). It is built around a physical string model that runs on two physically dislocated computers and uses the network latency between the two computers as a delay line in the sound engine. There are two human performers involved, one in each physical location, performing on a "shared" instrument. This instrument can be seen as a development of the concept of "cross-adaptive" performance (Brandtsegg et al. 2018) in which the sound-producing actions performed on one instrument influence the timbre of another instrument. Such shared instruments are constructed around combinations of acoustic and electro-acoustic instruments. They use machine-based sound analysis methods to extract some sound features from one instrument applied to the other. Since the instrument is already based on a network—both conceptually and technically—they lend themselves well to telematic performance.

There are additional conceptual challenges when performing together in different locations. No rooms are the same, which means that the perceived ambiance of the spaces will differ. As Braasch (2009, 424) reflects, ". . . our organism is formed by interacting with its environment. From a phenomenological viewpoint, it becomes much easier to treat the telematic system as a unique environment instead of a copy of another place, which a rationalist would call the 'real' or 'physical' world."

That is why we talk about the MCT Portal as a "physical-virtual" environment. The performers are colocated in two (or more) physical rooms connected with audiovisual technologies. Conceptually speaking, everyone is located in one common physical-virtual room. If you are on the physical Oslo side of the MCT Portal, you are at the same time virtually present in the other connected physical rooms and the other way around. While setting up the space, I was always concerned about creating a feeling of "presence," of being "together" with people on the other side of the communication chain. Too often, video conferencing setups lead to the experience of "us" and "them." Rather, we should think about a common "we." This is something that differentiates telematic performance from other types of performance. It is not only about "streaming" image and sound from one location to another. It is about creating a shared physical-virtual performance environment. You know that you are in your own physical environment, but you also know that you are elsewhere. Sometimes you know the other physical space well; other times, you are virtually present in physical locations that you do not know. Even with high-quality audiovisual representations of the other physical space, there is a difference between being physically and virtually present.

Telematic performance has been a niche activity for a long time but has nowadays become commonplace. Better technology helps, but so does more knowledge about the positive sides of playing with people that are spatially dislocated from yourself. More than ten years ago, Oliveros et al. (2009, 2) argued "As the technology improves exponentially and ubiquitously then eventually there will be no reason not to perform music at a distance. Globalization gives us more reason. Making music together makes friends."

As Oliveros notes, telematic performance is a way of connecting with other people. It could even be seen as an "intercultural instrument" according to Alarcón Díaz et al. (2019). In the project INTIMAL, she explored telematics as an interface for "relational listening." Here the idea was to explore how performing together, over long distances, could be a way of connecting with others. Her group of focus was Colombian immigrant women in Europe. They are both dislocated relative to their original home country (Colombia) but also currently dislocated in their various new home countries around Europe. Her project's final telematic performance was performed between the cities of Oslo, Barcelona, and London. We spent a great deal of time testing out various communication types for the performance. There were performers and local audiences in each of the cities, so we could have created a "normal" multimedia-style telematic performance through transmitting and playing back sound and video in each location. However, we wanted to see whether it was possible to create separate and colocated experiences simultaneously. We ended up transmitting the voices of the performers, but these were only heard by the other performers. So all the performers heard each other, but each local audience only heard the voices of the performers in their physical room. The only common auditory element played back in the three sites were sonifications of the women's respiration. The performers wore wrist belts that captured their breathing. The data was sent as control messages to the other sites where it was sonified using abstract sound models and played back (Alarcón Díaz and Jensenius 2019). As such, the result was three independent yet connected performances. They happened simultaneously but in different locations, and the content was a mix of the abstract, sonified breathing, and site-specific, context-dependent performance elements. This was a very different telematic performance. It was not only about streaming audiovisual information from one place to another. Instead, it considered the local environment and context, with the performers being the "mediator" of the content from the other side.

Space and Time

Telematic performances call for an expanded model of how one considers space and time in music. I here think about "space" as the experience of a physical "place." In acoustic musicking, the perceptual space often relates to the physical construction of the place. The room acoustics shape the overall musical experience for both performers and perceivers. Concert halls and churches are places that afford particular experiences of space. But any room or outdoor location will have unique sonic characteristics. As discussed in chapter 7, some environments can even be thought of as having instrument-like qualities, such as the entrance of the Nationaltheatret train station in Oslo (figure 7.10).

The question about space and place is more complicated in the world of electro-acoustic musicking. It is common to build "dry" rooms and add artificial environmental sound through the use of reverb and other sound effects. Some electro-acoustic instruments also have built-in reverb units to add spatiality to the output sound. Together this creates a mix of different spatial layers of sound:

1. Physical resonance in the instrument
2. Artificial reverb in the instrument
3. Physical resonance in the room
4. Artificial room reverb

In the case of telematic performance, all (or some) of these layers may be transmitted to the other side. They will be mixed and played over speakers in a different physical room with its acoustic characteristics. Analyzing the complete sound chain and how it is perceived is a complicated task. I find it helpful to think about what Moore (2016) calls the "space in sound" and "space through sound." His approach, however, is coming from the perspective of composing. I am more interested in how we can think about space and place in a performance context.

Traditional musical instruments, from guitar to synthesizer, are objects that are physically colocated with the performer. In most cases, the spatial dislocation is minimal in acoustic instruments, and the time delay is short. Electro-acoustic instruments may have a larger spatial dislocation between action and sound. They may also have different time delays. As described above, a musician onstage may have a stage monitor close to the instrument or an in-ear headset. This reduces the time delay of the action–sound (or sound–sound) mapping for the musician. However, the audience may experience both spatial dislocation and time delay if the speakers in the PA system are far from the performer.

The spatiotemporal distances found in telematic performance are of an entirely different magnitude. Here questions about in-time and out-of-time performance are more prevalent. However, the difference between the physical dislocation may not be proportional to the experienced latency. In the MCT Portal, we have optimized both the audiovisual setup and the network. This makes it possible to have a latency down to around twenty milliseconds between the cities of Oslo and Trondheim (five hundred kilometers apart). Speaking into a microphone in Oslo will result in sound output from a speaker in Trondheim with less experienced latency than if you sit in the back of a large concert hall. Latency of around twenty milliseconds is also fine for performing most types of music together.

"Here" and "Now"

When is the "now" when performing telematically? The performance is happening in real time on both sides, but what is the reference time? Can we talk about a common "now"? This may be thought of as similar to the question of when lightning happens. There are multiple "nows" that refer to each other. When I perform in the MCT Portal, I have my experience of performing in time. I am also aware that the performer on the other side has a different in-time experience than myself. It is common that both performers feel "dragged" toward the other's time reference. So when performing telematically it is necessary to resist slowing down to follow each other's in-time experience.

In the MCT Portal, we try to create the experience of a physical-virtual "here," even though everyone involved knows that there are multiple "heres" and "theres." In my experience, it helps to keep the communication channels running continuously. Video conferencing systems are often based on the idea of "calling" someone else or "connecting" to a different room. This underlines the idea of a "here" and "there." Instead, by leaving two (or more) locations connected continuously, you create a sense of togetherness. If you know that audio and video is transmitted, you enter a room with a different attitude. You know that, when stepping into the room, you are part of a physical-virtual "here."

To connect my thinking about space to that of time, I will suggest two new terms: "in space" and "out of space." Like the terms "in time" and "out of time," these space terms also require "here" as a reference point. When performing telematically, my "here" and "now" is different from my coperformers' "here" and "now." At the same time, we will both have an understanding of the other being "out of space" and "out of time" with respect to ourselves. We can play together because we are aware of each other's reference points.

However, being aware may not be sufficient in all cases. If the spatiotemporal distance gets too large, it may be impossible to play together in time. Then it may be necessary to resolve to other types of performance modes in which timing is less important. The user may still feel connected to the others. Swarbrick et al. (2019) investigated people's experiences with live versus recorded music. They found that people enjoyed live music more, possibly because it increased anticipation and feelings of involvement for the audience.

IV

Affection

10

From Ivory to Silicone

"Can you really feel the difference between the pianos?" Nonmusician friends often found it strange that I queued up early in the morning to get one of my favorite rehearsal rooms at university. There are considerable differences between instruments. The keys, the sound, the smell, and how they are tuned all play a part in defining an instrument's "feel." This is part of the tacit knowledge of musicians. Explorations into subjective experiences may help fill in some gaps missed by empirical research of other's experiences. In this chapter, I will describe my own experiences of playing with some commercial instruments and interfaces. While it feels daunting to use myself as an example, I have been inspired by others that have taken similar approaches. Sudnow (1993, 2001) describes learning jazz piano as an adult. Aho (2016) tells the story of playing the Finnish kantele. Both of these are examples of approaching an instrument (and musical genre) as an adult. While they have been going into great detail about their journey of learning a particular instrument, I will provide a comparative perspective on playing multiple instruments.

A Techno-Somatic Approach

Music evokes our senses and plays with our emotions. Philosophers have grappled with the question of emotions in music for millennia. Susanne Langer (1957, 235) suggested that "music can *reveal* the nature of feelings with a detail and truth that language cannot approach." Many studies on affect and emotion in music focus on the music "itself." Less attention has been devoted to the emotional experience of making music. As Shaffer (1989) argues, music performance is not only about the mechanical production of notes. For a robot to play a Chopin waltz convincingly, it needs to have feelings about itself and the social context.

Music researchers generally do not spend much time thinking about instruments. Similarly, musical instrument researchers do not spend much time researching performance. I find this lack of interest in instruments and music performance strange. After all, instruments are essential to musicking. Also, if you ask musicians about their instruments, they could talk for days. They would most likely also become quite emotional during the conversation. Still, there is little research on emotions and affect related to instruments. What is it that makes us want to play specific instruments? Which features and qualities do these instruments have that others do not? Is it only the age and cost of an instrument that make it desirable to play? Or the opposite, why are some instruments less engaging than others?

It should be no surprise that I believe the closeness between action and sound is part of one's affection for an instrument. Of course, some of the basics of good instrument design should first be in place. Sullivan and Wanderley (2019) found that professional musicians favor "durability," "portability," and "ease of use" when choosing their instruments. These are all important factors, but I think it is critical also to consider what Roads (1996) describes as the "feel" in his classic *The Computer Music Tutorial*:

> Electronic input devices detach the control of sound from the need to power the sound; any one of dozens of input devices can control the same sound generator. This translates into musical flexibility. With electronic instruments, a single wind controller can create the low bass sounds as easily as the high soprano sounds. Creating extremely soft or loud sounds requires minimum effort since the control is electronic. Obviously, the detachment of sound control from sound production has a negative side—the reduction of the "feel" associated with producing a certain kind of sound.

These thoughts are equally valid today. The increased action–sound separation of new instruments has also decreased the feeling of *inter*action when one plays an instrument. Playing the trumpet, one feels the lips' vibration toward the mouthpiece. It is a highly sensory experience and tightly coupled to the resulting sound. The same is usually not the case when playing an electro-acoustic instrument. But how does one explain such a "feel" of playing an instrument? In the following, I will present what can be called a "techno-somatic" approach to borrow a term from Paine (2015). In my thinking, the meeting point between a user and the instrument can be broken down into three elements: the "look," the "sound," and the "touch." Together, these three elements constitute the "feel" of an instrument, as sketched in figure 10.1. We will, in the following, investigate these concepts in more detail.

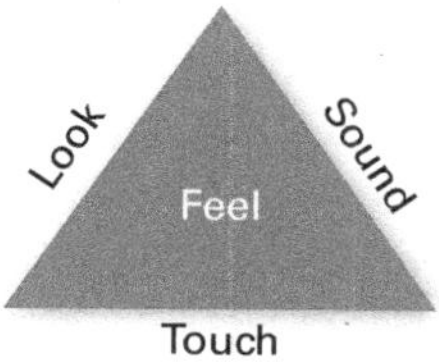

Figure 10.1
The "feel" of an instrument can be understood as a combination of its "look," "sound," and "touch."

The "Look"

Many people would probably argue that sound matters the most when "listening" to music. However, several studies have shown that vision matters more in music than we like to think. Tsay (2013) performed a series of experiments on conservatory competitions. In their self-reports, the expert judges said they focused most on sound in assessing the musicians. However, it turned out that visual information was the most important. Both novices and experts reliably identified the winners of the competitions from silent video recordings.

The visual importance of performance is reflected in the emphasis on stage setups, lighting, projections, extra video monitors, and so on in concert venues. Tickets are usually more expensive for positions where one can see the performers the best. For example, I prefer to sit on the center-left side of the hall when I go to piano concerts to maximize the possibility of seeing the finger action of the pianist. If going to a large stadium concert, I would instead position myself in the center to get a balanced view of both stage and monitors with close-ups of the performers.

The visual appearance of instruments also matters. The musical affordances are often conveyed through the instrument's physical construction. Thus, even before any sound has been made, the instrument's affordances have been "revealed" through sight. I think this matters to both performers and perceivers. Also, a designed object's color, shape, form, texture, and name tell something about who the object is made for and for what purpose. Most acoustic instruments have standardized shapes and colors, although there are some examples of violins painted red or pianos covered in silver. Such visual "hacking" is usually part of an artist's branding.

Electro-acoustic instruments are less standardized in their visual appearance, although many are based on keyboard designs or the "knobs-and-sliders" paradigm described in chapter 8. Many music technology devices also have a "techy" look. Jawad (2020) investigated how the color, form, and name of different music technologies affected peoples' gendered perception of the device. She selected nine example devices for an online gender assignment study and found that most were classified as male or neutral. Few were characterized as female. These findings correspond with general observations about the male domination in the field of music technology (Essl 2003; Born and Devine 2015; Gadir 2017).

Many traditional instruments are constructed so that it is possible to see the performance actions of the musician. That is not always the case with electro-acoustic instruments. For example, "knobs-and-sliders" controllers are usually not very audience friendly. That may be one reason that several musicians have started to project close-up videos of their performance actions so that the audience can see what is going on. The same is the case in laptop performances. Live coders often project their screen for the audience to avoid being accused of "reading email" on stage. That way, they can show how they are working with the sound-producing code. Other laptop performers have added visual elements to their performances (for example, projections of photos or abstract animations). There are several examples of outstanding audiovisual performances. However, I have been to performances with no apparent relationship between the visual and sonic elements. The visuals appear to have been added without much thought. Such audiovisual "dissonances" may ruin a performance. In some cases, it might be better to leave out the visuals. In fact, some performances of "tape music" are played in the dark. Then the audience is forced to focus on only listening.

The "Sound"

The sound of an instrument is essential for how it feels, both to the performer and the perceiver. There are many levels of meaning to consider. One can think of the sound metaphorically ("a scary sound"), generally ("the sound of a synth"), related to the device ("an 808 sound"), or related to an artist ("the sound of Björk"). From a perceptual point of view, one can argue that the most important is the experienced sound ("the sound of a synthesizer connected to a medium-sized PA system in a small club with a semidry reverb").

Many instruments are claimed to have "unique" sound characteristics, and one of the strongest examples of this may be found within the world of violins. Here the old Italian violins, particularly the ones made by Stradivari, have an exceptional reputation. Just mentioning that someone plays a Stradivarius will make people sharpen their senses. The question is whether the fame of an instrument also primes the way the instrument is experienced. Fritz et al. (2012) called in for a double-blind test in which a group of twenty-one soloists evaluated six violins, three old ones (Stradivari and Guarneri del Gesu) and three new ones. They found that the new violins were repeatedly preferred above the old ones. The study received much attention and also criticism about the experimental setup. Therefore Levitin (2014) set up a new experiment to improve the experimental design on many of the criticized points. Now they had ten expert performers try six old and six new high-quality violins in two different venues. They came to the same conclusion; the new violins were systematically preferred over the old ones. In a third experiment, they followed up by checking the instruments' perceived quality by audience members (Fritz et al. 2017). Again, the new violins were preferred to the old ones. While these findings caused harm, the researchers' main argument was not to devalue the old violins but to explain that new violins are equally good or better. I think the findings from these violin studies are interesting because they show that knowledge of a brand may get in the way of the instruments' sonic qualities. Without the effect of the brand name, it boils down to the instruments' qualities.

Many electro-acoustic instruments have been built around specific sonic qualities. While only experts can discern the exact model names, many would probably recognize "the sound of the 80s" based on some synthesizer sounds. In some cases, companies may also brand their "sound" as different from others. For example, some might say that a Moog "sounds" different than a Roland synthesizer. That may be true, but there may be a share of corporate interests or ownership bias involved.

The "Touch"

The third axis in my instrument "feel" triangle is "touch." The sense of touch is often broken down into three parts: cutaneous, kinesthetic, and haptic (Weiner 2003, 148). The cutaneous system refers to the sensations in the skin, the tactile experience of the object one touches. The kinesthetic system refers to sensory input from the muscles, tendons, and joints. The haptic system combines the cutaneous and kinesthetic systems and is focused on actively touching and being touched. To generalize, we may say that the tactility of an interface relates to what we feel with our skin, such as the surface material. The haptic experience is related to how the surface moves, the vibrations we feel in our body, or the pressure exerted on our body from the interface itself. Together, the combination of tactile and haptic experiences is part of what Wang (2018) calls the "resistance" of the interface.

The choice of material is essential for the overall tactile experience of an instrument. For example, think about the difference between touching wood, metal, or plastic. Even thinking about touching one of these surfaces will bring up mental images of how the materials feel. A metal object's flat and cold surface is undoubtedly different from the warm, textured wood feeling. The coating of the material also plays a role here. For example, the wood of an old scratched guitar is different from a shiny, new one. The size and form can also influence the way the instrument feels. The slides on my two trumpets are slightly different, making it necessary to adjust the hand position a little when I switch between them. These are tiny details, but they are still part of the "feel" of each instrument.

Although there are variations in tactile experiences between acoustic instruments, they are much more significant in electro-acoustic instruments. MIDI controllers come in all sizes, shapes, and materials: from the smallest, lightest controllers made of plastic to sturdy metal-based controllers with wooden elements. The latter would typically be the heaviest ones and would probably be perceived as more "expensive." However, why? Because of the touch? Or weight?

Weight is a particularly tricky topic. In one way, one would assume that a heavy instrument is more sturdy. Some studies show how an object's weight leads to an increased feeling of monetary value (Jostmann et al. 2009). On the other hand, we have been accustomed to smaller and lighter, yet more capable, devices through the digital revolution. A device's weight may also fool us. The plastic remote control for my new TV has a metal piece inside. The extra weight makes it sit better in hand. However, I would have preferred if the whole remote had been made of more sturdy materials. When it comes to instruments, heavier is not always better. If I can choose between two devices with different sizes and weights, I may be tempted to go for the lighter and smaller if it fulfills my needs. For musicians carrying their instruments around, size and weight are critical factors.

Up until now, we have primarily considered "static" touch properties of instruments. However, dynamic haptic feedback is also crucial for interaction. You get haptic feedback for "free" in acoustic instruments that the musician is holding or touching. However, as discussed in chapter 7, a larger action–sound separation will also lead to less haptic feedback. In electro-acoustic instruments, there may not be much haptic experience at all. Things are changing, though, and there has been a growing interest in haptic research over the last decades. One of my favorite examples is the haptic FireFader by Berdahl and Kontogeorgakopoulos (2013). This one-dimensional fader allows for creating powerful interactions based on providing haptic feedback with both a high spatial resolution and a high speed. Combining this interface with a physical string model makes it possible to feel like you are touching a "real" string. The FireFader provides only one-dimensional haptic feedback, but multipoint vibrotactile feedback solutions also exist. Musical haptics has up until now been a marginal research field but is quickly emerging (Papetti and Saitis 2018). Successful research prototypes have shown that haptics is a viable path for the future of electro-acoustic instrument design. However, engineering difficulties and costs have hindered the development. The growing interest in haptics in other domains, particularly in the game industry, may help develop better actuator technologies and reduce the cost of integrating them into devices.

From Acoustic to Hybrid Pianos

Keyboard-like musical instruments have a long and rich history (Keislar 1987), and keyboard-based music controllers have been dominating electro-acoustic instrument designs for decades. Therefore it makes sense to start an analysis of commercial instruments from the perspective of the piano. I will use my former piano experience when going through different keyboard-based instruments, from my childhood acoustic piano to new electro-acoustic instruments.

Acoustic Pianos

In my childhood home, we had an old upright acoustic piano (figure 10.2). It had already lived a long life when I started playing on it and was completely worn out when I moved out of the house. I still have an embodied memory of sitting on the quirky chair and touching the keys on that piano. The ivory keys had a rich, physical texture. Many had scratches, and some even had cracks on the sides. Based on the recommendation of a piano tuner, the keys were replaced after some years. While the new, shiny, white plastic keys were an improvement in general, I remember that I missed the feel of the old ones. The tactile sensation of the old keys was terrific, although their unevenness became a problem when I started playing more rapidly.

The felt on the hammers was also quite worn, and as I got older and started improvising more, I thought that the sound coming from the instrument was too dull. Sometimes when I was home alone, I removed the front plate to get more "punch" in the sound. This made the piano much

Figure 10.2
Practicing on the old piano in my childhood home.

louder, which was my primary motivation, but I also enjoyed watching the hammers move while playing. Already at that age, I was fascinated by the action–sound coupling of the hammer mechanism. Watching the hammers made me feel more connected with the instrument since I could see the sound-producing action.

The piano has a larger action–sound separation than what you find in embodied, tactile, and tool-based instruments. Still, the piano is a great first instrument. It has what Wessel and Wright (2002) call a "low entry fee." Pressing a key will result in sound right away, and the logical layout makes it easy to start exploring melodic and harmonic relationships. A drawback of this "easy-to-get-started" factor is that pianists do not develop the same sensitivity to pitch—a keypress results in sound. You can modify the velocity of the keypress, which will alter the loudness of the tone, but you cannot change the tuning. If the tone is out of tune, you have to get used to it and wait until the next time the piano tuner comes around. When I started playing the guitar at the age of twelve, I realized that I had been entirely ignorant of tuning. I had to work hard to incorporate a sensitivity to pitch. In contrast, those that play wind and string instruments spend a great deal of time learning to produce sound and intonate well.

On the positive side, playing the piano helps one understand the basics of Western music theory: the relationship between notes, the distribution of notes in octaves, and chord structures. The drawback is that the piano effectively "locks" you into a Western music-theoretical paradigm and makes it difficult to experiment with nondiatonic tuning systems. My guitar teacher was a composer interested in alternative tuning systems, and he showed me how to tune the guitar differently and the effect this had on what I could play. I also learned to appreciate being in direct contact with the sound-producing elements with the guitar: touching the strings and feeling the guitar's vibrating body. Picking up the trumpet gave me an even closer physical sense of producing sound myself.

Instruments are different, and they influence the music you make with them. A benefit of playing the piano is that you become an "all-round" musician. Pianists often have good multitasking skills: creating a rhythmic pattern, balancing harmonies, and playing melodies. However, as I discovered when I started studying music, the drummers will always be better at rhythms, and the wind players will always be better at melodies. On the other hand, the pianists will team up with the guitarists in being good at harmony. The guitarists usually have a better ear, though, since they are used to tuning.

The instrument(s) you learn shape your musical expertise and experience. Some people say that the instrument they play does not matter, that they have a musical idea in their mind and can play it on any instrument. For me, it is the opposite. The instrument I have at hand drives my musical creativity. Even when I try to play the same tune on another instrument, the result differs. The instruments have different qualities, and they afford other musical content and expression. To me, these possibilities and constraints of different instruments are attractive. The idiomaticity of various instruments gives them different musical affordances. That is why it is fun learning to play more instruments.

Many musicians are used to playing their own instruments, but pianists have to play on whatever instrument is available. The piano is one of a few instruments that musicians do not bring with them. There may be an economic benefit to not investing in a high-quality instrument on your own. The downside is that you rarely know what type of instrument you will play in a particular location. During my student days, I would chase for the rehearsal rooms with the best pianos. The pain of having to play on some of the worn-out or out-of-tune pianos was not tempting. The result was that I often played on two to three different pianos a day: my digital piano at home (more on that later), various upright pianos in the regular rehearsal rooms, and, when I was lucky, a grand piano in one of the large rehearsal rooms. Over the years, I have gained experiential knowledge of the qualities of various instruments. For example, playing an upright Yamaha piano with a "sharp" sound quality and heavy keyboard touch is undoubtedly different from playing a soft-touched Steinway grand piano. But even similar instruments from the same manufacturer differ. Acoustic pianos are unique instruments. Their differences may be minor, but they matter. Different touch and tone affect the musical result.

One of my favorite piano records—Keith Jarrett's *Köln* Concert—is an example of how a particular piano instrument greatly influenced the musical result. Carr (1992) writes about a misunderstanding about the instrument to be used for the performance. The instrument planned to use in performance was never moved to the Köln opera house. Instead, a small rehearsal piano was placed on the stage. This instrument was "thin" in the upper registers and weak in the bass and had problems with the pedals. The mistake was discovered too late to make a replacement before the concert. Despite the difficulties, Keith Jarrett decided to play. The Köln Concert has become legendary. It stands out as quite different from his other solo improvisations, probably because he had to adjust his playing to the peculiarities of that particular piano.

Digital Keyboards

After playing acoustic piano for some years, I got my first digital keyboard at around ten. It was a basic three-octave keyboard with nine instrument banks and some basic drum patterns. It had a speaker built in, so I would classify it as a complete electro-acoustic instrument. I still remember how strange it felt to touch the keys. At the time, I still played with the old ivory keys on the acoustic piano. The light keyboard keys felt slippery, and they gave no resistance compared to the acoustic piano. The disappointment with the keys quickly turned into the joy of exploring the keyboard's sonic possibilities. I particularly enjoyed playing with the different organ sounds. On the other hand, the "strings" never sounded like string instruments, and I was equally disappointed with the "wind" and "brass" sounds. However, the built-in drum machine was inspiring. The beats were basic, but they allowed for making up more complete songs. This was my first time exploring what I today would call a music maker.

In high school, I became interested in exploring the possibilities of digital pianos, including making multitrack MIDI recordings. I saved up money from a summer job and bought a Roland RD-600 digital stage piano. I had read through all the reviews I could find in various music magazines and found it to be the best option given my budget. All the reviews had focused on the quality of the keys and the built-in sounds. What I had not thought about, which has resonated with me ever since, was that the piano did not have any built-in speakers. I guess it was evident to "everyone" that such a stage piano does not make any sound on its own. Connecting it to my hi-fi system worked, but playing over two relatively small speakers on the other side of the room was not a satisfactory experience. I ended up investing in a pair of good headphones to enjoy the high-quality sound. My regrets eventually ended, and I have had many good musical experiences with the RD-600. The keys were the best you could find on a digital piano at the time. They were undoubtedly a little more lightweight than on an acoustic piano, but this was easily compensated by the possibilities of changing between different sounds.

The RD-600 has been with me ever since. The lack of speakers still bothers me. I usually play it with headphones in my office, but it is more of a struggle to play something for others. Then I have to connect it to my small PA system to get sound. It is doable, but it keeps reminding me about the unnecessarily large action–sound separation. The counterargument is that adding some small speakers into the RD-600 would be unnecessary, given that it was designed for stage usage. However, even on stage, some local speakers in the device could have been valuable. They could be used for warming up or to give some direct feedback to the performer. So the primary advice I give to everyone who wants to buy a digital piano is to get one with built-in speakers. That is, they should buy an instrument, not only a controller or keyboard. Fortunately, many digital pianos sold these days have speakers built in, although stage pianos and "synthesizers" typically ship without.

Hybrid Pianos

I always dreamed about owning my own grand piano. For a long time, this was an unrealistic dream. Good grand pianos are expensive and require a lot of space. After moving to a new apartment some years ago, the dream could finally become a reality. The challenge, then, was to decide on an instrument. I ended up with a Yamaha N2 "hybrid piano." This came as a big surprise to my surroundings, but also to myself. A few years earlier, I would never have thought to invest in a digital grand piano. My three arguments for buying a digital piano were that it takes up less space than an acoustic grand piano, it does not need tuning, and the sound level can be adjusted. The latter is perhaps the most important when you live in an apartment. One of the reasons I have been able to play much over the last few years is because I could turn down the volume so as not to disturb neighbors or wake up my daughters.

Before deciding what instrument to purchase, I spent time trying out all sorts of instruments in piano shops. Looking for acoustic instruments is challenging; they all have different characteristics. New instruments from the large manufacturers often have a "standardized" sound, but less "soul." On the other hand, used grand pianos from smaller manufacturers may be too "soulful." Parallel to trying acoustic grand pianos I also tested all the digital pianos I could find. None of the "normal" ones convinced me; the keys felt similar to the ones I already had on my RD-600. However, the "hybrid pianos" felt different. These pianos have the complete hammer mechanism that you find in an acoustic grand piano. Other digital pianos often have some weight attached to the keys to make them feel piano-like. However, they still do not feel like acoustic pianos. Hybrid pianos are based on the same full-balanced keys that you would find in a regular grand piano. They also have small actuators sitting in the frame, making the keys vibrate while playing. This haptic feedback makes a tremendous difference when playing. It feels like playing on an acoustic instrument.

The second selling point (for me) of these hybrid instruments is the sizeable built-in speaker system. Most digital pianos are lightweight in comparison. This makes them easy to move around, but it also means that the instrument can move while playing. Stage pianos on a folding stand are the worst, as they typically wiggle back and forth while playing. Digital pianos with a built-in stand also feel more flimsy than an acoustic instrument. On the other hand, hybrid pianos have the same type of instrument body as acoustic pianos. That means that they have space for a good sound system, including good amplifiers and speakers. They also distribute the sound spatially. When playing on the right side of the piano, the sound comes from the right side of the instrument. So there is a spatial relationship between action and sound. A prominent bass speaker also helps get a real sense of depth in the instrument when playing in the bass register.

My N2 looks fine, but the sound and touch make it "feel" just right. It is not the same as an acoustic grand piano, but it is remarkably similar. I sometimes regret not having an acoustic grand piano. Having a perfectly tuned instrument is practical, but I miss the "soulfulness" of a slightly out-of-tune piano. For some time, I thought that the hammer action and sound were essential parts of why the instrument felt so natural. However, then the vibrotactile actuator system broke down, and I discovered how much the haptic feedback influences the instrument's feel. I also learned that such instruments need attention over time. Fortunately, parts can be replaced to get them back in order. So the touch is important for how such an instrument feels. However, the essential feature is the ability to turn down the sound level at night. It does not help the feel, but it makes me play.

Electric Pianos

I have more experience with electro-acoustic pianos based on digital than analog circuitry. I have a couple of small analog synthesizers that I play with from time to time, but then I typically spend more time turning buttons than pressing keys. I think of them more as "synthesizers" than "pianos." But I have one strong memory with an "electric piano."

Some years ago, we rented a holiday apartment from a musician. In the corner of the apartment, partly hidden underneath a shelf and piles of books, stood a Wurlitzer electric piano. I had only briefly tested some old analog electro-acoustic pianos in the past. Now I had a chance to try one out for an entire week. The feel of playing on such an analog instrument is undoubtedly different from playing on a digital. People talk about the "character" of an instrument. To me, that boils down to its imperfections, whether in the look, the sound, or the touch. What made the Wurlitzer so enjoyable was that it was noisier than expected. It was slightly out of tune, and it had lots of scratches after being marked by many years' usage. It definitely looked and sounded more "authentic" than a digital piano.

I have fond memories of that week with the Wurlitzer, and it reminds me of the difference between analog and digital electronics. Analog electronics have some of those imperfections that you would also expect from acoustic instruments. That may be an asset in some cases but problematic in others. It is impossible to say that one instrument is better than another. All instruments have a different "feel," and they all afford other types of musicking.

Multidimensional Controllers

As described in chapter 6, MIDI Polyphonic Expression (MPE) has been developed to overcome some of the MIDI standard's limitations. New multidimensional controllers open for going beyond the impulsive sound-producing actions, predefined sound envelopes, and fixed amplitudes of traditional digital keyboard interfaces. These controllers are the result of several decades of research into alternative keyboard designs, such as Multiply (Moog 1982), a clavier optimized for curving pitch slides (Snell 1983), Rolky (Johnstone 1985), and Clavette (Fortuin 1995), to name a few. I call them "controllers" here since none contain speakers, and only a few contain a built-in sound engine. Over the last few decades, we have seen several such multidimensional controllers reach the market, including Continuum Fingerboard (Haken et al. 1998), Soundplane (Jones et al. 2009), Seaboard (Lamb and Robertson 2011), TouchKeys (McPherson 2012), and Linnstrument (Linn 2013).

Given my interest in keyboard-based instruments, I have followed the development of these new multidimensional controllers attentively. I will, in the following, reflect on my experiences with playing on several of these devices.

A Techno-Cognitive Approach

My description of pianos earlier in this chapter focused on explaining the feel of the instruments using a techno-somatic approach. This was operationalized through the experiential parameters of look, sound, and touch. I could have used a similar method also when analyzing the multidimensional controllers. However, I decided to conduct the analysis using a techno-cognitive approach to understand their inner workings better. This was inspired by the work of a former master's student of mine, Mads Bye, who used "dimension spaces" in an evaluation of multidimensional controllers (Bye 2018). He based his dimension spaces on the seven evaluation parameters proposed by Birnbaum et al. (2005). I will use a subset of these dimensions:

- **Plug-and-playability:** The time and effort it takes to make the keyboard produce sound.
- **Configurability:** The number of modification possibilities used prior to performing, such as setting up patches and loading samples.
- **Controllability:** The possibilities (and limitations) of the keyboard during the performance, such as the number of keys and the multidimensional control of each key.
- **Performability:** The subjective experience of how the instrument works in a real-world musical setting.

The two first dimensions deal with what happens before performing. The plug-and-playability is essential when moving an instrument around. Sometimes time is not a limitation; other times, it is critical. Sometimes I have chosen to play with a different instrumental setup because I know there would be little time to set up before a performance. The plug-and-playability does not only relate to time. Complexity may also be an issue. Some keyboards may rely on a particular hardware and software combination to work. Then it may be easier to choose another more flexible one.

The configurability dimension also relates to setting up a device, but typically after it is already functioning. Selecting presets and loading samples may be seen as analogous to tuning an acoustic instrument. It typically happens before one plays or in between songs. With complex electroacoustic setups, rehearsing what should happen between sets is critical. This could be seen as a type of non-real-time performance activity.

The controllability dimension may or may not overlap with the configurability dimension. The main difference is that the controllability dimension describes what happens during a performance. In general, this may be the various types of continuous controls, such as pressing the keys on a keyboard instrument. But it may also include loading samples or loading presets on the fly.

Both the configurability and controllability dimensions can be thought of as describing the degrees of freedom of a device, albeit from different temporal perspectives (before and during a performance). In chapter 4, I defined the degrees of freedom as the number of independent movement variables in a mechanical system. This variable is relevant when considering the control dimensions of the device, which is a techno-cognitive term describing the possible usage of a device. Typically, more degrees of freedom of a system may correlate with a higher "cognitive load" (Sweller 1994).

The number of degrees of freedom of a system is often challenging to operationalize. For example, how many degrees of freedom are there in a regular eighty-eight-note MIDI keyboard? There are eighty-eight individual keys to choose between, so one may argue that it has eighty-eight degrees of freedom. This makes sense if you think about the number of moving parts. However, each of the keys sends two control messages (pitch and velocity). Does this mean that you have 196 degrees of freedom? If we consider the usage of the controller instead of the device's construction, we may think about the hand's motion. Sideways motion controls the pitch selection, and vertical motion controls the velocity. Thus, one could argue that the controller only has two degrees of freedom (pitch and velocity). We should, in any case, also add the number of independent buttons and knobs on the device, such as data from the pitch wheel and connected pedals.

The exact number of degrees of freedom is less relevant than the complexity it reveals: the number of different "things" that the performer needs to consider while playing. Also, the complexity of an instrument does not equate with the complexity of the performed music. Good performers can create complex music with an instrument with few degrees of freedom. Similarly, a too-complex instrument may not lead to any music at all if the performer cannot produce or control the sound. Thus, the optimal controllability of an instrument is often a sweet spot between its degrees of freedom and cognitive load.

Fagereng (2008) studied the performance setup of the Norwegian guitarist Eivind Aarset, a jazz musician known for working with many samplers and effects pedals. The study revealed that Aarset's arrangement increased in complexity over many years. Here complexity was defined as a combination of the number of devices, their degrees of freedom, and the cognitive load put on the performer. At some point, this complexity became counterproductive, and Aarset began scaling down the setup to (re)gain a level of control. In interviews, he confirmed that introducing constraints by removing gear and possibilities helped him be more creative in performance. So scoring high on configurability and controllability may, in many cases, lower an instrument's performability.

Many acoustic instruments can be thought of as having high plug-and-playability. For example, an acoustic guitar can be picked up and played (almost) right away. One may need to tune it, which could be seen as part of the configurability dimension. Still, a guitarist would be able to start playing within minutes of unpacking their instrument. The same is the case of a cellist, although a bit more preparation time is needed. First, you need to pick up the bow and tighten it by turning the tension screw. Next, you may need to apply rosin on the bow hairs. Then it is time to tune the strings. Brass instruments can often be played quickly, and installing and wetting the reed of woodwinds does not take very long either. When they are ready, these acoustic instruments can be played for a whole concert without interruption. Some (re)tuning may be necessary between songs, but it is usually relatively quick. Brass players need to empty the spit valve, and others may need to adjust some mechanics. Occasionally, longer breaks are necessary, such as when a string breaks or a reed needs replacement. Still, the general level of performability is high in acoustic instruments.

The traditional instruments have been optimized and standardized for centuries. As such, they can be seen as having found different sweet spots between the various dimensions. In the world of electro-acoustic instruments, things are different. Many instruments have been designed and prototyped over the years, fewer have been produced commercially, but only some have become "mass products." Thus, few electro-acoustic instruments—and here, I do not include hybrid instruments like the electric guitar—have been around long enough for performers to develop a career-long performance practice. A few professional musicians have specialized in playing early electro-acoustic instruments, such as the theremin and Ondes Martenot. There are also some new interfaces for musical expression that have been used over longer time periods, such as Michel Waisvisz's The Hands (Waisvisz 1985), Cléo Palacio-Quintin's Hyper-Flute (Palacio-Quintin 2008), Hans Leeuw's Electrumpet (Leeuw 2012), and Laetitia Sonami's Lady's Glove (Fiebrink and Sonami 2020).

One could argue that keyboard players specialize in playing on electric pianos and synthesizers. However, most keyboard players change their setups over time. Many now also include computers as part of their instrument. This increases the configurability and controllability, albeit at the cost of plug-and-playability. For example, computer musicians may arguably be the most "flexible" in terms of what can be achieved on stage, but this often comes at the cost of long setup times. Driver problems are less problematic today than a decade ago, but I still frequently run into connection issues when setting up laptops with sound cards and controllers. Additionally, software needs to be started, sound files loaded, and mappings set up between controller and sound engine. Furthermore, even when everything is set up, there is always the risk of things not working. This is particularly the case when working with wireless and battery-powered devices. In comparison, bringing a hardware synthesizer to a gig is more comfortable. Then it may only be necessary to plug a power cable into a wall socket and connect a sound cable to the mixer. Some digital electro-acoustic instruments may be ready to play in seconds. Many mobile phone apps score high on plug-and-playability. A laptop can, too, if used without any peripherals. In most cases, a laptop will also be more configurable and controllable than a mobile phone. Whether that leads to a higher level of performability is another question. Sometimes they do, other times not. In my experience, playing laptop-based setups often relies on particular patches and ready-made material. Thus, the higher level of configurability leads to a longer time needed to move between settings in performance.

One often talks about the "expressive potential" of instruments. But what does that entail? Dobrian and Koppelman (2006) questioned that many new digital controllers focus primarily on technical aspects. Less focus has been put into how we can make instruments that allow for expressive performance. Since then, there has been a growing body of research into using body control. This is an interesting line of work, and

I will present some of my explorations into body-based instruments in chapter 12. However, there is a misunderstanding that adding gestural control to an electro-acoustic will make it more expressive. That is built on the same misconception that adding more possibilities to a synthesizer will make it more performable. As Magnusson (2010) has argued, having endless possibilities may lead to a feeling of despair. He calls for the need to design constraints into new musical interfaces. Creativity blossoms if you work with a more limited instrument.

Interestingly, traditional acoustic instruments are full of constraints. In the mapping analyses by Kvifte (1989), most acoustic instruments have relatively few control dimensions. They also have a reasonably narrow action–sound palette. But why are such instruments then thought of as expressive? I would argue that expressivity is linked to a combination of an instrument's action–sound separation and the spatiotemporal distance between instrument and performer. You do not need a lot of options if the spatiotemporal resolution of the interaction is high. As such, all the multidimensional controllers in the following sections score higher than a traditional keyboard. They all allow for nuanced control possibilities, albeit in very different ways.

McPherson's TouchKeys

The first of the multidimensional controllers I will discuss in this chapter is TouchKeys by McPherson (2012) (see figure 10.3). TouchKeys is different from the other included keyboards in that it is not a controller on its own. Instead, it is a regular MIDI keyboard controller with a set of key-shaped sensors glued on top of the keys. I have decided to include it here because it is a relevant "bridge" between a traditional digital keyboard and the MPE interfaces we will consider next.

The TouchKeys sensors allow tracking the XY position on each key. The input from the sensors comes in addition to the MIDI controller values from the keyboard itself. Therefore, TouchKeys behaves like a standard MIDI keyboard plus multidimensional control information for each key. The controller's plug-and-playability is relatively low since the values come in through two separate sources (MIDI input and sensor data input). It takes some time to use the data meaningfully. In fairness, TouchKeys should be seen as a research device more than a commercial product, so users would expect to know how to work with the data.

The controller itself does not allow for many configurations, and the controllability is limited to the user's programming. However, TouchKeys shines when it comes to performability. It can be played with a

Figure 10.3
Playing with TouchKeys, Continuumini, and Roli Seaboard in the office. The devices are connected to an Onde loudspeaker.

standard keyboard playing technique and allows for continuous sensing on each key.

The richness of the continuous sensor data streams is daunting at first. I always dreamed about continuously controlling the sound on each key on a keyboard. However, I never expected it to be so challenging to use in practice. At first, I started playing with a standard piano technique and tried to add various effects. The challenge is that playing with impulsive actions does not work well when you have continuous control. It is necessary to adopt a different playing technique to get something meaningful out of the controller. I have found that the most interesting thing about TouchKeys is the ability to add vibrato and pitch shifts on individual keys.

To avoid continuous control of each tone, I have found it necessary to program small "wait" commands before the continuous control kicks in. That means I can still play impulsive piano actions and get normal piano-like sounds if I play rapidly. However, when staying on a key, the continuous control will begin after a split second. This allows for a combination of impulsive and continuous control based on the temporal performance context.

Pianists are used to dealing with discrete note values. You can press anywhere on the key and get the same tonal result. However, how does one handle the spatial (XY) sensor data? I have found it challenging to use absolute values when mapping continuous sensor data to sound parameters. Hitting a key in the same position without any reference point is impossible. For some sound parameters, such as a filter, it could work with using imprecise position control. However, for pitch shifting it makes more sense to calculate the relative distance from the initial pressure point. Still, it takes time to get used to thinking about both relative and absolute key positioning.

When I first approached the TouchKeys, I used a standard piano technique with pressing anything from four to eight keys simultaneously. I quickly realized that having continuous control on each key results in a much higher cognitive load. Even after some time, I still have problems using much more than a few keys in a controlled manner. Playing sustained tones on TouchKeys, you continuously need to consider the position and pressure level.

Another challenge is how to move between keys. It is possible to slide sideways on each key but impossible to slide between pressed and unpressed keys. This differentiates TouchKeys from some of the other "keyless" multidimensional controllers. TouchKeys' biggest strength may be that it allows for using a traditional keyboard technique. This may also be its main limitation since it effectively hinders utilizing the full potential of its sensing capabilities.

The touch of TouchKeys is quite different from what I am used to from pianos and keyboards. The TouchKeys sensors are made of plastic but have another coating that's different than regular keyboards. Each key is also perforated, making them feel stickier than other keyboard controllers.

TouchKeys was the first multidimensional controller I played, and it opened my eyes to the world of continuous keyboard control. I have had a great time exploring its possibilities and limitations. At first, I naively thought adding continuous control to all keys would make such a piano-like controller more expressive. I have had to adjust my expectations and understand why all the new MPE controllers have opted for designs that deviate from a traditional keyboard design.

Roli's Seaboard Grand

My first encounter with the Roli Seaboard was during the International Conference on New Interfaces for Musical Expression in Oslo in 2011. Lamb and Robertson (2011) presented a prototype of a silicone-based piano claviature and later commercialized the product. There are nowadays several different versions of the controller. I purchased one of the early Roli Seaboard Grand devices, a sixty-one-note controller that has later been discontinued (see figure 10.3). The Seaboard does not have individual keys. Instead, it is built around a continuous silicon surface with "bumps" indicating the keys. The softness of the surface makes for a different tactile experience. One almost gets the sense of "faux" haptic feedback when pressing hard into the surface. One compelling design feature is that the silicone material and sensing extend beyond the multidimensional keys. This allows for using the upper and lower part of the keyboard as long touch strips.

When it comes to plug-and-playability, the Seaboard Grand scores high. It does not have a built-in speaker, so you still need to connect it to headphones or an external speaker system. However, with a built-in synthesizer, you can start making sounds right away. New versions of the Seaboard do not have a built-in sound engine, which means that you rely on an external sound engine to produce sound. To me, this makes a difference. I connect the Seaboard to a laptop once in a while, but I play with its internal sound engine much more times. That is why the Seaboard is one of the alternate controllers I have played the most, and it is the one I typically showcase when I have visitors in my office.

The levels of controllability and configurability on the device itself are not exceptionally high. There is volume control, and it is possible to change between sound presets. Many varied presets make it possible to produce different sounds, most of which utilize the multidimensional control possibilities. Of course, connecting to software makes it possible to fully explore the potential of both the controller and the sound engine. I find it is interesting that the sound engine has been developed specifically to explore the multidimensional control possibilities. The Seaboard was developed before the MPE standard, and there were few multidimensional sound engines around. So it was necessary to think about both action and sound to realize the potential of this novel controller.

When I received the Seaboard, I thought the piano-like keyboard would make it easy to play. However, playing scales on the soft keys requires practice. This is even more evident when playing chords. Finger placement is easy, but intonation is difficult. At first, I thought that the possibility to slide between individual keys was exciting, but I have found it is easier to play with sliding-like actions above or below the keys. That extended area functions as a long touch strip. Then it is possible to get the sense of playing a sustained instrument rather than just pressing on/off on a controller. As for performability, one thing that helps when playing tonal music is the "pitch lock-in" feature on individual notes. This feature makes it easy to play in tune, but it is also a limitation if one wants to play more freely. There is a similar type of lock-in feature when sliding an octave. It is, of course, possible to program it otherwise (in software), but I have come to see these features as idiomatic to the device.

All in all, the Seaboard is fun to play. Being able to have multidimensional sound control is immediately rewarding. The lock-in features make it easy to get started and results in few surprising results or false triggering. Even though it is easy to get started, it is not an easy instrument to master. I have found that having a piano background only partially helps. Multidimensional control is quite different in practice. As for the feel, the Seaboard is unique. The metal casing is sturdy and feels solid, and the touch of the silicone surface is quite unlike anything else I have played. It also looks different, which adds to the package.

Madrona Labs's Soundplane

The Soundplane was described as a "force-sensitive surface for intimate control" when it was first presented by Jones et al. (2009). It has later been commercialized by Madrona Labs (see figure 10.4). The controller has 150 "pads" that respond to velocity, pressure, and two-dimensional position tracking. What makes this controller unique is the playing surface, which is made entirely out of walnut wood. This gives it a look, touch, and smell that resembles acoustic instruments.

The plug-and-playability is relatively low. The Soundplane has no inputs or outputs except for a USB port for communication and bus power. It relies on proprietary OSX software, which only handles the parsing of control data. So you need an additional sound engine to make it work. There are no onboard controls for alternative configuration modes, and the software editor also has limited alterations, so it also scores low on configurability and controllability.

Fortunately, the Soundplane shines when it comes to performability. In many ways, it is the opposite of the Seaboard. The flat control surface of the Soundplane invites continuous control between the "keys." There are small gaps between the keypads, but it is easy to slide between them. It is tricky to play individual notes, scales, and chords; the Soundplane works much better for timbral and textural control. Like with the TouchKeys and Seaboard, I find it challenging to handle more than a couple of continuous tones simultaneously.

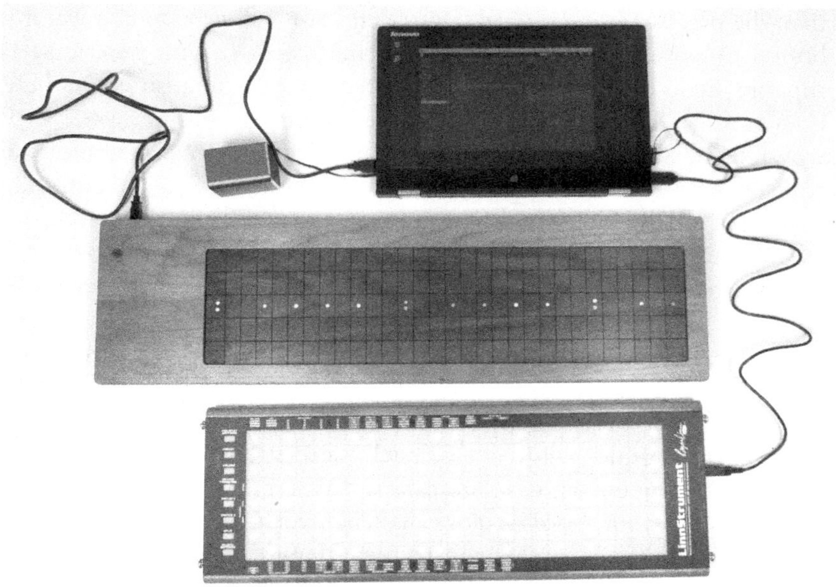

Figure 10.4
The Linnstrument (bottom) and Soundplane (middle) are examples of multidimensional grid-based controllers. They rely on a computer running a sound engine and a speaker to make sound.

What works well with the Soundplane is the controller's high spatial and temporal resolution and low latency. This makes it possible to explore sonic features with a high level of detail. The Soundplane is my controller of choice if I want to test a sophisticated sound engine. I often tell my students that even the most basic sound engine can come "alive" with a high-quality controller. The high spatiotemporal resolution of the Soundplane affords inspiring sonic results.

Roger Linn's Linnstrument

The Linnstrument 128 is an MPE-compatible controller with 128 RGB back-lit rubber pads (see figure 10.4). The pads respond to velocity, pressure, two-dimensional position, and release velocity. The controller connects to the computer via USB and is bus powered. As such, it resembles the Soundplane in many ways. It is about the same size and also has a grid-like construction. However, there are many differences.

The fact that the Linnstrument can function as a regular MIDI device right away makes the plug-and-playability of the controller much higher than the previously mentioned controllers. You still need to connect it to both a sound engine and speakers, but that is easier with generic MIDI support. Enabling the MPE mode requires extra fiddling, so I would give it an average "score" on the plug-and-playability dimension.

While the Soundplane relies entirely on software for configuration and control, Linnstrument has many more options built into the controller itself. It is possible to change the tuning of the pads, split the layout, and use the built-in arpeggiator, step sequencer, "strum" mode, and so on from the controller itself. It takes time to understand how these settings work. However, once learned, you can change a lot of settings on the controller.

At first, I thought that the RGB-lit plastic pads were a bit "cheesy." For example, the Soundplane's wood surface feels so much more organic in comparison. However, I have found that the slightly elevated keys make playing tones, scales, and chords much more straightforward. The fact that keys light up in octave relationships makes it easy to jump around on the surface. Playing chords is more challenging. Unlike the highly sensitive Soundplane, the pads on the Linnstrument require a certain level of touch to be activated. The keypads feel a little stiff, but the response is consistent. Since there is a clear gap between keypads, the controller feels more key based than continuous. It is possible to slide between keys, but it does not feel natural. That said, the Linnstrument scores high on performability. It is easy to navigate even after a short practice time, and it feels musical right from the start.

Haken Audio's Continuumini

The Continuumini is a controller produced by Haken Audio (see figure 10.3). It is the "little sister" of the piano-sized Continuum Fingerboard but otherwise shares many of the same properties. It is made of metal and has a flat felt-like playing surface. It is perceived as a sturdy device with a tactile feel.

Like the Seaboard Grand, the Continuumini has a built-in synthesizer. There are no speakers, but the addition of a sound engine makes it possible to play without a computer. This also leads to a "zero-latency" feel when playing. The Continuumini has few options for configuration beyond changing presets. It allows for changing octaves on the device, which is practical for such a small control surface, but otherwise relies on configuration options in the software editor.

As opposed to the other controllers discussed here, the playing surface on the Continuumini is so tiny that it primarily affords you to play with one or two fingers at a time. At first, I thought that this would be a limitation. However, given the high spatiotemporal resolution of the device, more focus goes into shaping individual tones through continuous control.

Continuumini's control surface is connected to mechanical springs, so the whole surface moves up and down when playing. This results in some mechanical "noise," which I find charming given the controller's otherwise digital nature. The springs also give some resistance, so even though there is no digital haptic feedback, the controller has mechanical feedback.

When it comes to performability, the Continuumini is a rewarding controller. I am still working through the myriads of presets on the device and have found many smart mappings. For example, several of the presets allow for playing both with impulsive and sustained actions. This is a similar type of mapping that I ended up making for the TouchKeys. The result is that it is possible to play impulsive-like sounds with impulsive actions while at the same time being able to stop in one position and work with sustained control. The addition of various after-touch features that start when one lifts a finger from the surface is also cleverly programmed. All in all, the device feels very musical.

Comparing Apples and Pears

The above-mentioned multidimensional controllers are the first of a new generation of commercial music technology devices. The advent of the MPE standard means that many more will probably be produced in the coming years. While the existing devices share some characteristics, there are also many differences. That is refreshing, given that MIDI controllers have become fairly standardized over the years. I will not dwell too much on discussing details that will quickly be obsolete. I am more interested in reflecting on some of the techno-somatic properties of such multidimensional controllers.

Let us start by considering the "look" of the devices. While plastic has taken over as the primary construction material in many commercial products in the last decades, I have been positively surprised to see products manufactured in metal and wood. However, using such materials is costly. It should be noted that these controllers are exclusive products and are not priced to meet the mass market. Still, I hope that we can get to a point at which solid construction becomes viable again. After all, considering the time it takes to master a musical instrument, they should be built to last. This is also important from a climate perspective. It is not sustainable to continuously make new products that break after a short period of usage. In addition to the more robust construction, working toward standardization of communication protocols and software-based updates may also help keep controllers alive longer than they have been over the past decades.

The materials used in a controller also influence their "touch." Again, I have been positively surprised by the variety of materials used in the playing surfaces in these multidimensional controllers: rubber, silicone, wood, and felt. This leads to a different playing experience than you usually get with plastic-based controllers. It certainly feels liberating and gives each of the controllers its unique tactile signature. Some of them even have a sense of haptic feedback, such as Soundplane's silicone-based keys and Continuumini's spring-based surface. The Linnstrument does not have such haptic feedback, but the light feedback on each key works well. Since the playing surface is otherwise just a grid of white buttons, it helps that the buttons light up to show adjacent tones and octaves.

Since none of the controllers have built-in speakers, and only two of them have built-in sound engines, it is impossible to compare their "sound." It should be clear by now that I favor instruments that produce sound on their own. Seaboard and Continuumini have built-in synthesizers that are optimized for their control capabilities. I can use these controllers also with generic sound engines, but I enjoy exploring the mappings developed by the instrument makers. This gives these particular controllers more of a musical identity. There is also a practical component to this. The plug-and-playability is much higher when you just need to turn on a device to make sound. There is no need to turn on a laptop, connect the device, and start up one or more software packages to make everything work. Having a built-in sound engine also helps to lower the latency in the system. Even though they do not have built-in speakers, they feel more like complete instruments. I see that I play the Seaboard and Continuumini more than the other controllers.

Given that all the devices allow for controlling multidimensional sound engines, it has been fascinating to explore the same sound engines with each device. This has made me realize how much a controller influences sound production. You would imagine that playing the same sound engine with similar controllers would result in the same sonic and musical results. However, as I have spent time getting to know each of the controllers, I have realized that they afford different types of sonic control. For example, the Soundplane lends itself better to continuous control of timbres and textures, while the Linnstrument is the one to reach for when playing traditional melodies and harmonies.

Another general impression is that the multidimensional controllers provide excellent control over pitch and timbre. Paradoxically, this continuous control also limits the number of tones that can be played simultaneously. Before I started playing such devices, I thought multidimensional control would allow me to play like a pianist with timbral superpowers. Instead, I have found myself hitting the techno-cognitive "glass ceiling." Miller (1956) elegantly summarized the limits of human information processing capacity as "the magical number seven, plus or minus two." On a regular piano, I can easily play harmonic progressions with seven tones at a time. However, on these multidimensional controllers, I typically play one or a few tones at a time. I have found it impossible to play full chords with individual intonation on each tone. This is something that musicians playing on continuous-control acoustic instruments could have quickly informed me. After all, they spend an entire life playing one—or only a few—tones at a time.

11

Unconventional Instruments

"Does it have to look like an instrument?" one of my students asked. It was the first time I was teaching a course in sound programming, and I had given them the assignment to design and build an electro-acoustic instrument. Most of the students had plans for making software-based designs, possibly hooking up a MIDI controller for the interaction. The question sparked off an interesting discussion about the visual appearance of instruments. What does an instrument really look like? Is it possible to design an instrument that does not look like an instrument? These questions have been with me ever since. Over the years, I have created various instruments. Mainly electro-acoustic, but also some acoustic. In recent years, I have become more interested in hybrid devices that bridge between acoustic and electro-acoustic instruments. In this chapter, I will present some of my unconventional instruments and controllers. These are designed to be different. Similar to the impossible instruments mentioned in chapter 7, unconventional instruments are not only sound makers. Equally important is how they challenge our thinking about what an instrument can be.

An Action–Sound Design Philosophy

There is no universal recipe for the design of new instruments; there are just too many variables involved. It is not difficult to make an instrument. Anyone can construct an instrument with whatever materials they have around. Music technology educators have long experience in teaching students how to build simple acoustic and electro-acoustic instruments. We probably also have many different strategies for how to make students succeed in developing new instruments.

In many ways, this book summarizes my instrument design philosophy. Some may find it strange that I spend so much time dealing with nontechnological topics. When I began teaching music technology courses, my approach was very technology focused. I taught the students about the nuts and bolts of signal processing, sensor interfaces, and programming. I ensured that they mastered sound synthesis before we moved on to control the sound engines. It was a structured approach to electro-acoustic instrument development, but it was not very inspiring. It was also an utterly disembodied approach.

Over time, I gradually developed what I have called an action–sound design philosophy (Jensenius 2013a). This is based on the embodied cognition theory we looked at in chapter 2. In particular, I have been fascinated by how the human body interacts with the world through objects. Papanek (1985) has written about the need to design for the "real world." One approach to doing that is by embracing the affordance theory of Gibson (1977) and its application to design theory by Norman (2013). In his books on the design of everyday things, Norman focuses on creating intuitive objects. They should feel "natural" to the user, whether it is constructing a door handle, a teapot, or computer software. Later on, he has also called for the need for more "emotional design" (Norman 2004), which in many ways overlaps with thoughts on "affective computing" (Picard 1997). A red line through all these design approaches is that new technologies need to be embodied and human focused. When working with computer-based systems, it is easy to end up with technology-oriented solutions. That is because we are developing based on the computer's logic. However, how can we design for humans?

Wang (2018) suggests that we should aim for creating "sublime" designs. One of his strategies to reach such sublimity is by creating "playful" designs. This will make users physically and emotionally engaged. This resonates with how Norman (2004) suggests that we need to take three levels of human processing into account when creating designs:

- Visceral: the automatic, prewired layer
- Behavioral: the brain processes that control everyday behavior
- Reflective: the contemplative part of the brain

A challenge is how to achieve all such design ideas when "anything" is possible, particularly with computer-based instruments. It is like bringing someone into a large workshop and asking them to make "anything." That may be a more difficult task than giving them a small piece of paper, a stencil, and three colored pencils. Limiting peoples' possibilities fosters creativity. Magnusson (2009) argues that electro-acoustic instruments should be developed based on a "constraint-based design." Too many possibilities are daunting to a user. Therefore, designing constraints into devices helps trigger action. These constraints may, in turn, spark off novel usage. As Magnusson (2018) writes, the instruments may "teach, adapt, explain, direct, suggest, entice."

In his "laws of simplicity," Maeda (2006, 89) suggested that "[s]implicity is about subtracting the obvious, and adding the meaningful." However, what does simplicity mean when it comes to musical instrument design? Many acoustic instruments require years of practice to master. Hence, only a few musicians will ever reach a level of virtuosity on their instruments. Still, many people manage to get started with an acoustic instrument quickly. For example, most will produce sound the first time they pick up a guitar. Many also learn to play a few chords and a scale in a relatively short time. However, to move from novice to master is, indeed, a long journey.

Wessel and Wright (2001) argued that simple controllers lead to simple interaction. Such devices may feel more like a toy than an instrument, which may discourage continued musical exploration. Instead, the aim should be to create instruments that have a "low entry fee with no ceiling on virtuosity" (Wright 2002). They later regretted that phrase, suggesting instead to focus on designing "high-ceiling toys" (Wright 2017).

In my thinking, one approach to designing such high-ceiling toys is by increasing the spatiotemporal resolution of the controllers. Adding multidimensional control is another way of increasing the interaction possibilities of an interface. However, as discussed in chapter 10, adding more possibilities may also increase the cognitive load. Then it may help to leverage the idiomatic nature of existing instruments, building on what feels "natural" to play on a particular device. As Tahiroglu et al. (2018) have argued, "A gesture vocabulary that is idiomatic to the new musical instrument establishes a set of playing techniques for that instrument, which is, in turn, an essential component of developing a performance practice."

Several other researchers have argued for the need to consider idiomaticity more carefully, in both acoustic (Aho 2016; De Souza 2017) and electro-acoustic (Frisk 2008; McPherson and Tahiroglu 2020) musicking. However, idiomaticity may come in different forms and is highly dependent on the mappings between action and sound. If the mappings work well, an instrument is typically also experienced as both intuitive and simple to use.

Simplicity can take different forms whether we think about instruments as sound makers or music makers. For example, accompaniment keyboards focus on maximizing the musical output while reducing the interaction complexity. Such instruments often target beginners and build on the idiomaticity of the piano keyboard. At the same time, these instruments embody much musical knowledge through built-in sound and music engines. Through the touch of a button, the user can play an accompaniment (primarily chords and rhythms) following the melody played by the user. As such, the musical output is much more complex than what would be achievable with a traditional sound maker. Nevertheless, the user controls the musical output, but not beyond the musical structures preprogrammed into the instrument.

We have just seen the beginning of such semiautomatic instruments. In the future, I am sure that different types of machine musicianship will be the norm. Some may find this problematic, but almost twenty years ago Machover (2004, 171) argued for the benefits of such an active approach to music:

> What if we could unlock the expressive mysteries of music first, before learning the technical foundations, if we could help young people—and others—fall in love with the joys of music first, subsequently demanding deeper knowledge once they were "hooked"?

In the Toy Symphony and other projects at the MIT Media Lab, Machover has explored semiautomatic instruments that allow for active musicking. The users, often children, perform complex music with simple, toy-like controllers. The idea is that the instruments should be immediately easy to use yet provide rich musical experiences. The instruments may not allow for developing virtuosity, but they create excitement through musicking. That has also been one of the goals of my instrument designs.

CheapStick

My explorations into new instrument design came through software development. Then one typically puts together existing hardware and focuses on programming mappings between controllers and sound engines. CheapStick was my first attempt at creating a hardware controller (Jensenius et al. 2006). Nowadays, there are numerous affordable sensor interface solutions and embedded computing platforms around. At the time, there were no affordable solutions in place. While visiting the Input Devices and Music Interaction Laboratory (IDMIL) at McGill University, I teamed up with Mark Marshall and Rodolphe Koehly. We set out to create a complete and easy-to-build music controller for "ten dollars."

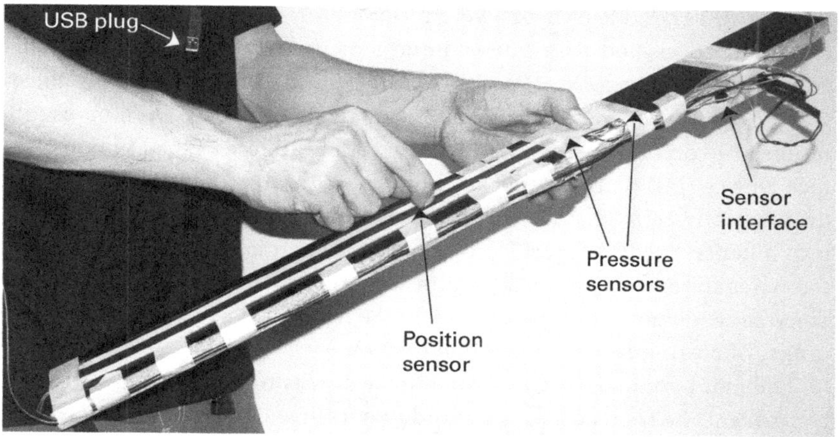

Figure 11.1
The CheapStick is a self-contained music controller built for around twenty-five dollars.

The result was slightly more expensive, but still lived up to its name: the Cheapstick (see figure 11.1).

The project started from ongoing experimentation with using the conductive capacity of various types of inexpensive materials. Position sensors and force-sensing resistors (FSRs) are commonly used sensor types in experimental instrument designs (Marshall and Wanderley 2006). However, there are several drawbacks to such sensors. Besides the cost, commercial sensors come in standard sizes and with a fixed resistance. These standards may not fit with a controller's design. Thus, it is necessary to adapt the design to the sensors' shape, material, and resistance rather than the other way around. Furthermore, many commercial sensors are made of plastic, which is not ideal from an environmental perspective. The plastic also gives a suboptimal tactile experience. Covering a small sensor with padding material may increase the tactile experience but also reduce its resolution.

Instead of adding lots of sensors to increase the controller's response, we decided to use the padding itself as a sensor. Inspired by the "shoe controller" of Paradiso et al. (1997), my colleague Rodolphe Koehly set out to explore various types of pressure and bend sensors from conductive ink, adhesive, rubber, tape, elastics, and paper. All of these materials are readily available in many different sizes. That makes it possible to be more creative in the interface design.

The final CheapStick prototype consisted of three pressure sensors made from conductive paper and a position sensor made from videotape (see figure 11.1). The sensors were mounted on a leftover part of a bookshelf and connected to a sensor interface created by "hacking" a game controller (Collins 2006, 195–200). Game controllers are generally cheap, stable, small, and readily available. Furthermore, they do not usually require any drivers or software installs, as they adhere to the Human Interface Device (HID) protocol. They also do not require any external power, drawing power from the USB port. A regular game pad can be turned into a generic sensor interface by removing the plastic shield, cutting off the wires to the built-in sensors, and soldering on new sensor connectors. Since these devices comply with a standard zero- to five-volt sensor input range, one can interface most sensor types.

The main goal of the Cheapstick project was to develop an affordable controller. The total cost was around twenty-five dollars, which was more than our target but much cheaper than building a similar controller with commercially available interfaces and sensors. The result was a simple, handheld controller that could be plugged directly into a computer. I developed the mapping software MultiControl to accompany the controller. This software allowed for easy mapping to any OSC or MIDI-based sound module, and it also served as inspiration for the Gesture Description Interchange Format described in chapter 6. Rodolphe Koehly continued his research on the use of different types of accessible materials and how to improve the consistency and reliability of the response of such sensors (Koehly 2011).

I still have the CheapStick sitting on a shelf in my office. It has not been used for years, but it reminds me of the need to consider affordability when designing new technologies. Many people do not have the financial means to buy commercially available controllers. There is much potential in reusing existing technologies and materials. This is an example of what Freed (2012) calls "sustainable instrument design." The Cheapstick showed that it is possible to create homemade sensors with similar, or even better, response and tactility than commercial sensors, and at a fraction of the price.

Music Balls

My explorations into designing simple and affordable music controllers continued with what would become a whole series of "music balls" (Jensenius and Voldsund 2012). The starting point was to challenge the idea that music technologies need to be colorless, have square corners, and contain many buttons and knobs. Many commercial controllers also have "techy" names in the form of meaningless numbers and letters. Things have improved slightly over the years, but the music technology industry is remarkably conservative. Commercial music technology products also still appear to be made by and for male engineers (Jawad 2020).

When thinking about alternatives to rectangular devices, the sphere came to mind. Creating ball-like controllers also encourages playful usage. Throughout the years, I have developed many music balls together with my students. These music balls have all had different visual designs, technical solutions, and action–sound mappings. Still, there have been some standard design features. They have been:

- **Round:** designed around the sphere as a geometrical form
- **Affordable:** built with cheap technologies, often leftover parts from other projects or items that we found lying around
- **Basic:** built with only a few sensors, hence constrained in their action potential
- **Limited:** designed with one action–sound mapping in mind, breaking with the idea that music technologies should be able to do "anything."

Most of the music balls were built with some existing ball or spherical shape as a starting point. In particular, I found that toy balls worked well for creating handheld music balls. They are durable and often manufactured in various colors and with different surface designs (see figure 11.2). Larger music balls were built from boat fenders and buoys, which are durable and readily available in many different sizes and shapes.

The first music balls I made focused on amplifying acoustic sounds. I would cut the ball and stuff them with various sound-producing objects, such as paper, peas, plastic, or steel wool. The challenge was to find materials that were both durable and nice sounding. For example, newspaper sheets have a crisp sound when squeezed but contract quickly. Then plastic sponges work better since they expand quickly after being squeezed. The amplification of sound was done by placing microphones inside the balls. For some balls, I used contact microphones placed in the middle of the sounding material. Other times I used elements from broken dynamic microphones. Ripping out the microphone membrane and soldering on a new cable is an easy way of reusing the valuable parts. My favorite music balls are the ones that use dynamic microphone elements wrapped in steel wool. They produce clear and crisp sounds when amplified.

I have had much fun building and performing such music balls with students. Acoustic music balls can be assembled quickly in class, and they can be directly connected to a speaker or guitar amplifier. Many students

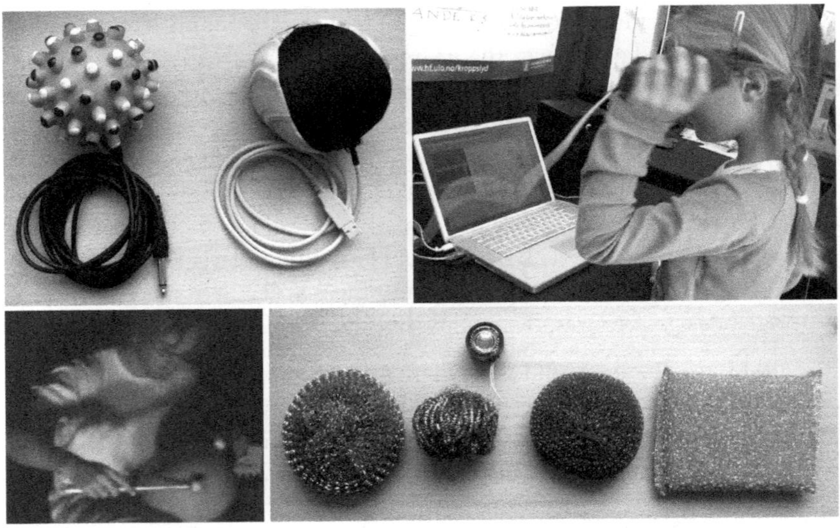

Figure 11.2
One acoustic and one electro-acoustic music ball (upper left). An electro-acoustic music ball in use at a research fair (upper right). Performance with a hybrid music ball (bottom left). A dynamic microphone element and various materials used inside of music balls (bottom right).

have also explored using various sound effects in the chain, such as guitar pedals. Other times, we have connected the balls to a computer to extract sound features from the sound signal. Then the microphone signal can be used as a "sensor" to control digital sound synthesis. Combining these approaches is possible, effectively making the music ball into a hybrid device combining its audible sound with digital sound effects.

In most cases, I have focused on having only one microphone or one sensor per music ball. Sensor-based balls are often built around an accelerometer or a pressure sensor placed inside a ball. Having only one sensor per ball makes them easy and cheap to produce. There are also fewer things to break. And, if a ball breaks, you have other balls available. Working with only one sensor per ball limits the interaction possibilities. An affordance-based interaction design also helps create more intuitive mappings. For example, an accelerometer-based music ball typically lends itself to creating sounds based on shaking the ball. On the other hand, a ball with a flex sensor inside allows for squeezing. At first, many students say they need more sensors to create interesting sonic results. However, when I tell them that they could also include time as a variable, they often develop creative sound interaction designs. There is a tendency to think about static mappings. Once you create a mapping, that is what you have. However, one of the potentials of electro-acoustic instruments is that you are not bound to the laws of nature. Creating mappings that change over time can be seen as a "gamification" strategy that may create excitement if done well. The trick is often to start with an ecologically meaningful mapping, which may evolve into something more creative. This is also a way of developing music makers instead of sound makers.

Big Buoy

After creating a range of small and simple music balls, I was challenged to make a big one for a research fair. This led to the creation of Big Buoy, named so because we built it from a large ship buoy (see figure 11.3). Here we ended up breaking with the initial "rule" of only using one or a few sensors per ball. One reason to include multiple sensors was that we developed Big Buoy as a multiperson instrument. Strangely enough, multiperson instruments are a rarity. Most musical instruments are built as a single-person device, although some—of which the piano may be the most common—can be played by two. Contemporary multiperson instrument

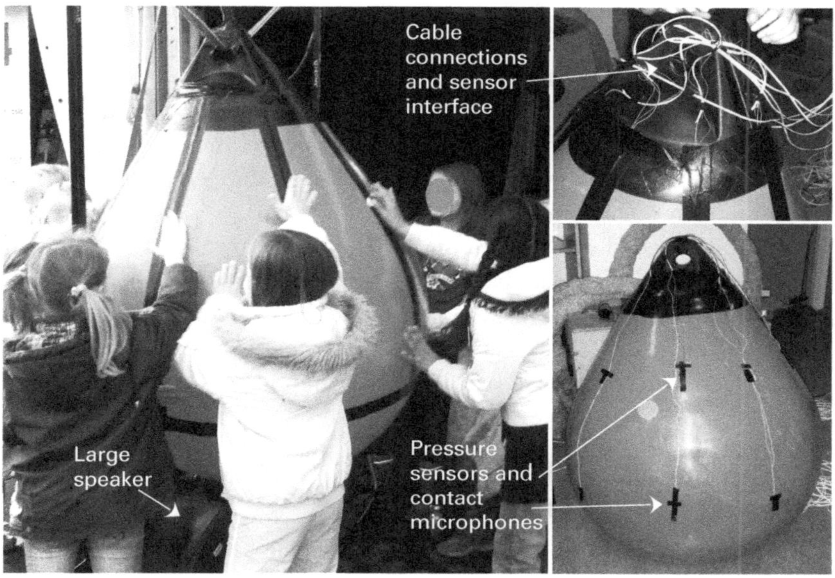

Figure 11.3
The Big Buoy is a multiuser instrument built with contact microphones and force-sensing resistors.

designs can often be found as museum installations rather than on stages. However, there are some exceptions, including the two-person instrument Tooka (Fels and Vogt 2002) and the multiperson reacTable (Jordà et al. 2005).

Big Buoy was conceived of as a multiperson instrument. It was built so that it would not work well with less than four people playing concurrently. This called for more sensors, but we still tried to limit ourselves to only a few sensing modalities. Building on knowledge from the hybrid music balls, we placed strips of contact microphones and pressure sensors on the sides of the ball. This made it possible to pick up sound-producing actions with the microphones and use the sensors to control sound effects. Since the ball was hanging freely, we placed a 3D accelerometer at the top to pick up motion patterns.

Before I started making music balls, I had not reflected much on the importance of local sound. However, when playing with multiple music balls, we quickly realized that it was necessary to separate the sound sources physically. The first attempts at connecting multiple music balls to a large PA system failed. The lack of local and directional sound was confusing both for performers and perceivers. We had therefore started to use small speakers close to each ball. We also experimented with placing speaker elements inside the balls, but this turned out to be tricky to realize given time constraints and limited engineering skills. For Big Buoy, we initially thought about embedding speakers inside of the ball. However, given the buoy's soft material, this turned out to be problematic from both a sound projection and interaction perspective. Instead, we placed a large PA speaker underneath the ball, which was large enough to produce low-frequency sounds matching the size of the ball.

We developed the initial mappings for Big Buoy based on previous experience playing music balls in various concert settings. However, when setting it up as an installation in Oslo's main shopping street, we quickly realized it was necessary to alter the mappings. For many coming by, our carefully developed action–sound mappings did not work well. The ball afforded rougher actions than we had anticipated. We, therefore, hastily adjusted the sound design to match the interaction mode. We also discovered that there was some time-of-day differences in the interaction patterns. More children were interacting with the ball in the mornings, while more adults came by in the afternoons. Hence, we also made two sets of mappings, with one being more "child-like." In its first installation, thousands of people interacted with the ball. Many people commented positively about the unusual shape and size of the instrument.

ADHD Ball

An exciting possibility appeared when researchers at the Norwegian Institute of Public Health asked us to develop a music ball for clinical experiments on children with "attention deficit hyperactivity disorder" (ADHD). They were running a large-scale, longitudinal study and wanted to engage the children while at the same time capturing information about their behavior. The idea was to make a music ball that would trigger different sound and light stimuli and collect data to study the childrens' response patterns. The ADHD ball was yet another exploration into the development of a musical instrument not meant to be an "instrument." Rather, it was presented as a toy that the kids could play with when they came to their clinical testing at the hospital.

To allow for rough usage, we decided to develop a scaled-down version of Big Buoy. We used a similar buoy type, stripes of force-sensing resistors (FSRs) fastened to the surface and an accelerometer at the top (see figure 11.4). The buoy was padded with foam, and a heavy-duty party balloon was stretched around the buoy to serve as a protective outer skin. The ball was nice looking and had a rubber-like tactile feel.

The ball was popular with the kids and was in daily use for more than a year. Even though we had done our best to protect the electronics, we had

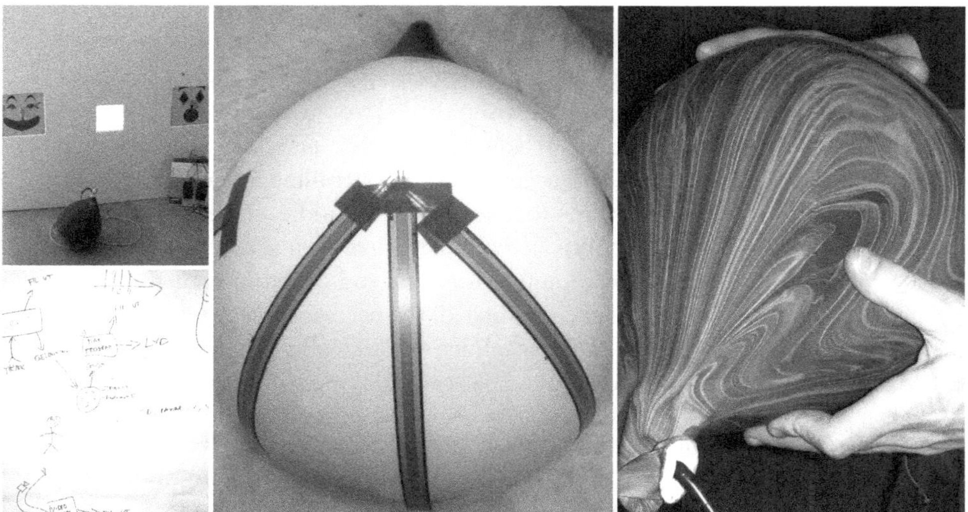

Figure 11.4
Pictures from the design process (bottom left), sensor construction (middle), padding (right), and installation of the ADHD ball (upper left).

to replace parts regularly. That is probably what to expect when you have continuous interaction over extended periods. Over time we also made slight adjustments to the action–sound mappings. Here we continued our experimentation with time-based mappings. This was done by changing sound material over time as if moving from one game level to another. If one would stop using the ball, the mappings would "reset" to their initial state, and the user would need to start playing again to reach the new levels. Such "gamification" helped to keep the childrens' attention.

Music Troll

It is not ball shaped, but the Music Troll is yet another unconventional instrument designed with simplicity and playfulness in mind (see figure 11.5). The instrument was built as a "many-headed" device and named after the Scandinavian fairytale "troll" creature. The idea was to create a collaborative interface based on different music balls. Like the construction of the small music balls, each head of the Music Troll was based on only one sensor type. In the end, we deviated slightly from the ball concept and, instead, made four heads of different shapes and sizes.

The four heads covered different tonal and timbral ranges that fit well together. The first head was built with a dynamic microphone element surrounded by steel wool. The sound's envelope was used to control the speed and velocity of a voice sample played backward, creating a strange, voice-like sound. The second head was cone shaped, with five "fingers" created by fastening strips to a plastic container with a dynamic microphone element inside. The sound from this microphone was amplified and compressed heavily to create a "squeaky" sound. The third head was ball shaped and contained a USB accelerometer controlling a percussive physical sound model. Finally, the fourth head was paddle shaped with a long bend sensor inside that controlled a sweep filter on a synthetic sound.

Each of the heads was mounted to metal rods fixed to a custom-built wooden box. This box also contained all the necessary electronics, including a computer, sound card, amplifier, and speakers. There were four speakers, one on each side of the box, each playing sound from one head. That way, people could more easily hear their contribution when playing on the Music Troll collaboratively.

Similar to Big Buoy, the Music Troll was also built as a multiperson instrument. It was set up as an installation piece in public venues in the Oslo area several times. Also, this instrument received rough usage by thousands of people. As opposed to the open-ended interaction design of

Figure 11.5
Performing with the Music Troll in Oslo Concert Hall in October 2006.

Big Buoy, the Music Troll had four separate heads with distinct musical sound designs. As such, it functioned primarily as a music maker. People could control the sound on each head to some extent but within relatively narrow borders. This worked well to grab peoples' attention, but it did not encourage extensive musical explorations. I also did a few stage performances with the Music Troll (see figure 11.5). Then it was necessary to create different mappings, allowing for more nuanced control and the possibility of moving from head to head. However, performing alone was challenging. I had to move around the box to reach the heads, and the lack of coperformers led to a suboptimal experience.

From Prototype to Prototype

I have learned some lessons after designing and building many unconventional controllers and instruments over the years. First, I have learned that it can be rewarding to develop simple instruments with intuitive mappings. It is always tempting to add more sensors and features when developing instruments, but this may come at the cost of user engagement. Being forced to work with only a few sensing modalities leads to creative solutions. It also makes it possible to develop more devices. The music balls are not the most advanced instruments, but they are fun sound makers. Some of them are also quite elaborate music makers. One of my goals has been to show that making simple yet musically valuable technologies is possible. Just in the same way that one says "the best camera is the one that you bring along," we could say that the best instrument is the one you have at hand. Many of my students got an eye-opener when they understood that they could build an instrument themselves.

In my classes, students have built their own music balls from leftover materials: plastic, leather, wood, and fabric. The aim has been to show that it is possible to reuse materials. This is something we need to think more about in the future from an environmental perspective. The same is the case with repurposing technologies, such as reusing microphone parts. It is both cheaper and more sustainable. Reusing parts also led to more varied-looking technologies. We have also succeeded in creating "non-techy" instruments. Hiding the cables and electronics has been an essential part of why the devices have appealed to broader audiences.

Stability and durability are the two most significant drawbacks of all the devices presented in this chapter. Except for the ADHD ball, one could say that we have moved from prototype to prototype. Jack et al. (2020) argue that there is a difference between an instrument prototype—a functioning but unfinished product—and a "research product" that has the qualities of a finalized instrument. A research product is different from a commercial product in that it is made to produce knowledge. On the other hand, a commercial product results from an elaborate engineering process targeted at consumers. I think it makes sense to add yet another category to the list: "research prototypes." An instrument prototype may be made as a first step toward making a complete musical instrument. On the other hand, the primary function of a research prototype is to generate knowledge about the design process and how instruments can produce a particular type of musical experience. My experimentation with unconventional instrument designs was an eye-opener to understanding more about what an instrument can be. It also paved the way for "air instruments," which we will look at in the next chapter.

12

Performing in the Air

"Can you be my air guitar coach?" I still get calls from people that want to improve their air guitar skills, typically right before the annual Air Guitar Championships in Oulu, Finland. This is a fascinating competition in which people compete in mimicking the sound-producing actions of guitarists combined with extrovert communicative gestures. The reason people call me is that we researched air instrument performance (Godøy et al. 2006b). Our primary research focus was air piano, but we also did pilot studies on air drumming and air guitar performance. The latter was picked up by a journalist who named me "Dr. Air Guitar" in the national media when I defended my dissertation in 2008. The fact is that I am no good at playing air guitar. However, I have, over the years, developed several "air instruments" in which one can produce sound by moving in the air. This chapter looks at different types of air instruments and how they can be used in performance.

Some Considerations When Designing Air Instruments

Air instruments are built differently, but what they have in common is that the user controls sound by moving the body in the air. I will focus on three types of air instruments that I have explored over the years:

Touchless air instruments: This is the "purest" form of air performance, without holding anything in the hands. The theremin is the classic example of such an instrument, played by moving the hands within the field of the instrument's two antennas.

Object-based air instruments: Such performances are based on holding a motion-sensing device in the hands. Some devices have buttons and knobs, so they can also be used as traditional touch-based instruments.

Muscle-based air instruments: These devices use the muscle tension of the performer to create sound. The most popular have been muscle-sensing armbands, but sensors can also be placed in other body parts.

Many of the air instruments we have developed at the University of Oslo have come out of our curiosity about how people move while listening to musical sound. This led to the experiments mentioned in chapter 4, such as the "air performance," "free dance," and "sound-tracing" studies. The idea behind these studies was to understand more about people's music cognition by studying how they spontaneously move to musical stimuli. One finding from these studies is that air performance is ubiquitous. As would be expected, people with musical training are, in general, better at carrying out such tasks. For example, they are better at picking up on more layers in complex musical stimuli. We were more surprised by the capabilities of people with little or no musical training. Even people who claimed to be "tone deaf" or "unmusical" could follow various musical features in the air with one or more body parts. In essence, people are more musical than they believe.

If moving *to* sound in the air is something everyone can do, what happens if we use this embodied music capacity to create new instruments? Several people have asked me to create an actual air guitar, a setup where you can play guitar-like sounds by moving in the air. There are already a couple of such examples (Karjalainen et al. 2006; Crawford and Fastenow 2009), although both of those setups were based on (partly) handheld sensing. However, just like I do not find it particularly interesting to (re)create traditional acoustic instruments with digital technology, I am equally uninterested in creating "air versions" of traditional instruments. Instead, I am curious about new instrument concepts based on people moving in the air.

As briefly surveyed in chapter 4, many technologies allow for capturing information about human body motion. It is impossible to say that one system works better than another; they all have positive and negative sides. The instruments I will present in this chapter are mainly based on two sensing types: inertial sensor-based units and infrared camera-based systems. The former is more flexible and embeddable. Inertial measurement units are small and can be placed anywhere on the body. This has made them popular in commercial products, which has helped lower the cost of such sensors. Camera-based systems are complex and costly in comparison. A multicamera setup is less flexible to start with, and the setup is sensitive to light pollution and reflections. However, despite these drawbacks, camera-based systems can capture the exact location of an object in space with submillimeter accuracy and precision. So we have found such systems to be excellent for prototyping and some occasional performances. Knowledge from such exploration can later be implemented using other sensing technologies.

One of the challenges of working with motion tracking systems is the massive amounts of data they provide. A stream of full-body motion data typically contains up to thirty three-dimensional (sometimes six-dimensional) markers at rates above 100 Hz. It is impossible to map such data directly to a sound engine; multiple levels of data reduction and analysis techniques have to be applied. One approach is to identify "key postures" in the continuous motion stream based on a particular configuration of markers in time and space. If one uses a full-body motion capture system, this could be based on a kinematic model of the body. However, other constellations of markers or sensors could also work. Such posture recognition can be seen as a kind of "spatial thresholding," looking at the spatial relationships between markers independent of time. This is an effective way to reduce the continuous stream of data since everything other than the identified postures will be discarded. One of the first examples of using camera-based motion tracking in real-time musical applications was a project by Gang Qian et al. (2004) in which they used such static postures to drive the interactive system. This was done by dividing the body into ten "rigid objects" and using the object's angular relations as features in a pattern recognition system.

Posture-based tracking is a good starting point, but to utilize the potential of a continuous stream of motion data, one needs to find solutions for handling the temporal aspects of the stream. One approach is to perform some kind of action recognition based on "temporal segmentation." Here the idea is to consider the temporal unfolding of motion. A straightforward approach is to calculate the running "quantity of motion" (QoM). This is a "perceptual value," meaning that it is calculated differently depending on the motion capture system in use. The point is to extract a feature that tells something about how much motion there is at any point in time. This feature can then be used to segment the continuous motion stream into action segments based on a threshold. This way of thresholding the signal keeps more information than a purely posture-based analysis yet discards everything under the given threshold. Several of the first motion capture–based instruments used this approach, such as in the music–dance performances created by Dobrian and Bevilacqua (2003).

Action-based tracking is challenging. A "touchless action" can be defined as an action performed in the air without any haptic and tactile response. Everyone can do it, but it is challenging to figure out how to extract such an action from a continuous stream of motion data. One approach is to do something similar to how a MIDI keyboard works. A keyboard passes on information about the finger's velocity during the key attack. The prefix and suffix of the action are irrelevant; the velocity of the attack is captured. One can do something similar by looking at the velocity of an object in space. The velocity will be zero when you switch direction while moving. Then one can use the energy profile leading up to the direction switch to control sound playback. This is an example of relative action recognition. The action can be performed anywhere; it is the energy profile that defines the interaction. A different approach is to use an absolute spatial reference. Just as a physical instrument is located in one position, one could define an invisible "line" that one can "hit." This is more in line with how a keyboard works.

Finally, a more high-level approach is to consider the meaning-bearing components of actions: the gestures. Many claim that they work with gesture-based interaction. However, according to my definitions, most such systems are motion or action based. But there are some examples of systems that aim at extracting meaningful features from actions. For instance, Antonio Camurri and colleagues at the University of Genova have developed the EyesWeb computer vision software for extracting expressive features from video recordings (Camurri et al. 2005). This software aims at capturing some of the quality of the motion in question and using these data in the mapping process to musical features. One challenge, then, is that musicians are generally not trained at performing such expressive gestures. Most musicians spend time improving sound-producing actions on a physical instrument. When it comes to performing expressive gestures in the air, this is the domain of dancers. However, dancers often have little training in creating sounds. So gesture-based interaction is in a somewhat gray area between music and dance performance.

Laserdance

My first exploration into "air performance" was in the music–dance project Laserdance. I met choreographer Mia Habib at the Norwegian Center for Technology in Music and the Arts (Notam) in Oslo in 2001 and developed the sound interaction design for one of her pieces. In

Figure 12.1
The red laser beam split the stage in two during the performance of Laserdance (top). The technical setup was based on a custom-built Notam sensor interface and an infrared sensor (bottom left). The interactive sound design was implemented in Max/MSP (bottom right).

Laserdance, the scenography was based on a red laser beam splitting the stage floor (see figure 12.1), and the choreography explored this divided space. Passing the laser beam would trigger a sound, and the dancers explored moving in and out of the beam during the performance.

The technical setup was based around an infrared sensor pointed in the same direction as the laser beam and connected to a custom-built sensor interface. This interface sent out a MIDI control message whenever the dancer moved into the beam. The control message triggered playback of a synthetic sound based on frequency and amplitude modulation in a Max patch. As such, all parts of the action–sound mapping were quite basic: the control signal, the sound engine, and the mapping between action and sound.

Despite the simplicity of the design, the setup proved remarkably effective. This was at a time when most dancers had no prior experience with interactive music systems. So being able to control sound themselves on stage was an eye-opener. The dancers moved around on the dark stage and produced sound when "breaking" the laser beam. They did not control the sound beyond starting and stopping the playback, but that was enough to make a meaningful performance. The audience commented on the effectful combination of visuals and sound. The project showed that often simpler is better.

Soniperforma

Camera-based interactive systems often rely on the extraction of features from the video frames. For example, the quantity of motion of an object can be estimated by calculating the number of pixels that change between frames. This number can then be mapped to a sound feature. An entirely different approach is to sonify the video, that is, create sound directly from the incoming data. This is the method used in the instrument Soniperforma (Jensenius and Godøy 2013). Here I developed a technique I called sonomotiongram, based on sonifying motiongrams. A motiongram is a compact visual representation of motion images, made by "squeezing" together motion images to create a motion timeline (Jensenius 2013c). The sonomotiongram technique plays with the idea that motiongrams are visually similar to sonograms (spectrograms of audio files). Sonograms and motiongrams are completely different visualization methods, but they may sometimes look alike. Therefore, I was curious about what would happen if I converted a motiongram to sound through an "inverse FFT" technique (see figure 12.2).

When converting an image into sound, one challenge is creating mappings between spatial (timeless) image features and nonspatial (temporal) audio features. In Soniperforma, data from the *Y*-axis of the motiongram is mapped to sound frequency. Data from the *X*-axis is used to control changes over time. Such a mapping from image to sound has been used several times in the past. An early example is the Pattern Playback machine

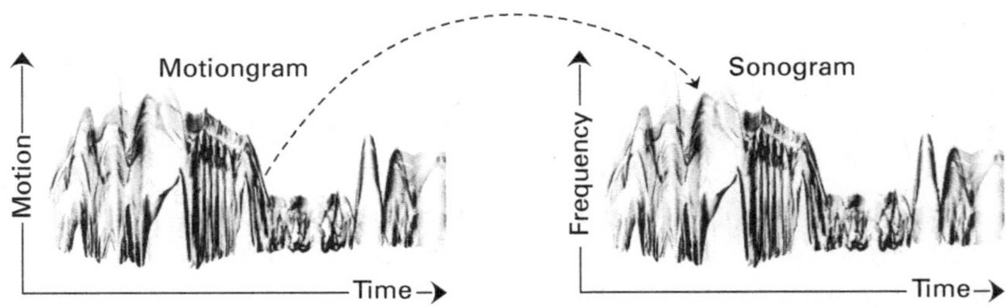

Figure 12.2
A sketch of the idea of sonifying motion through treating motiongram data as if it were a sonogram.

built by a group of speech researchers in the late 1940s (Cooper et al. 1951). This system made it possible to "draw" shapes that could be played back as sound. Iannis Xenakis's UPIC system from 1977 allowed for creating complex timbres by drawing with a digital pen on a computer screen (Marino et al. 1993). Wenger (1998) implemented similar ideas in the Metasynth software for sonifying images and photos. These are examples of sonifying still images or moving images with a still character. There have also been experiments with sonifying moving images. For example, Norman McLaren created a series of short films in which he drew with a pen directly on the soundtrack of a 35-mm film strip (Jordan 1953). In the 1970s, Erkki Kurenniemi developed the video-based musical instrument Dimi-O to control sound synthesis from a video camera (Ojanen et al. 2007).

The sonification technique implemented in Soniperforma is conceptually similar to the examples mentioned above. One of the challenges is the "flat" spatial representation. When time is running along one axis, there is only one dimension to represent the spatial properties. A motiongram will only display motion in one spatial dimension: a horizontal motiongram visualizes vertical motion. Reducing three-dimensional motion into a one-dimensional display is a limitation. Still, preserving temporality helps represent the motion's "shape." The result is an abstract yet connected auditory representation of the original motion.

As argued by Winters et al. (2012), sonifying body motion is not the same as creating musically meaningful mappings. A sonification aims to convey information about the data, not to create an aesthetically exciting sound. At first, I mainly thought of the sonomotiongram technique as helpful in video analysis. By converting a video file to sound, I could quickly listen through hours of video recordings at high speeds to detect deviations in motion patterns that are difficult to see with the naked eye. As it turned out, I found that the technique was also a good starting point for music performance.

The software-based instrument Soniperforma was developed as a patcher in the graphical music programming language Max using a combination of Jamoma modules (see figure 12.3). At the core of the instrument is the sonomotiongram algorithm, which converts the video stream to audio. As such, the mapping from motion to sound is fixed. However, I found that the sound could be controlled by modifying the incoming video. Changing the color, size, and orientation of the video image or applying visual effects would end up as sound effects. For example, a motion blur effect will end up as a sound delay. I have found this audiovisual

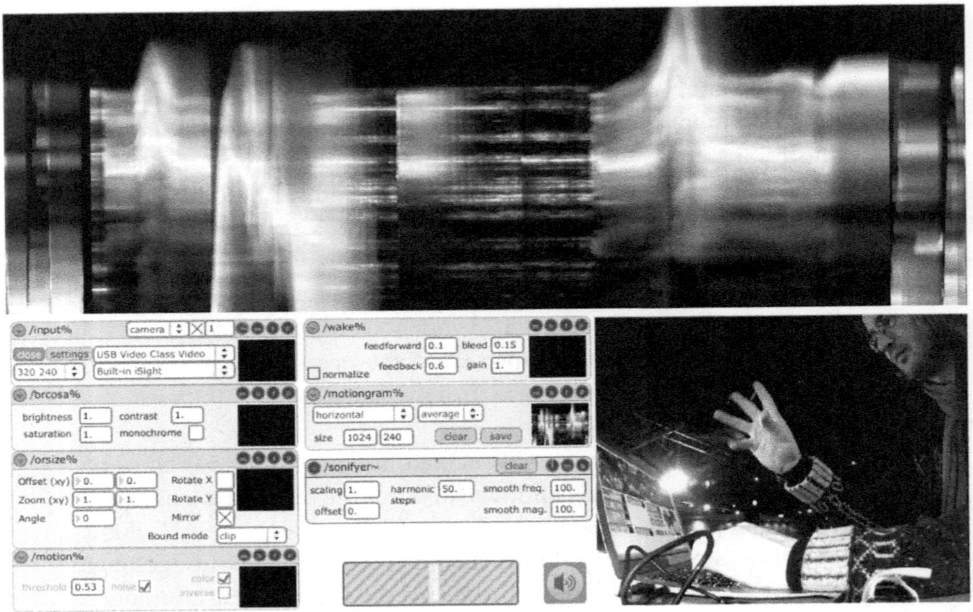

Figure 12.3
The performance environment of Soniperforma, showing the Jamoma modules used for controlling the incoming video signal (bottom left), the motiongram used as the basis for the sonification algorithm (top), and the incoming video (bottom right).

interaction creatively inspiring as a performer. Since I often project my screen during performances, it is also an audiovisual experience for the audience.

Its continuous one-dimensional interaction mode is both a strength and weakness in Soniperforma. For most physical instruments, it is possible to choose between sound-producing and non-sound-producing actions. For example, when playing the piano, only the keyboard actions produce sound. When playing with an instrument like Soniperforma, on the other hand, any motion in the image will show up in the motiongram and will be rendered as sound. Such a direct relationship between motion and sound is intuitive. One could even argue that it leads to a relatively small action–sound separation. Still, when performing with the instrument, I often wanted to turn the sound "on" and "off." One solution to this "always-on" problem was to position the camera so that I could move out of the image. I also explored placing the camera at a ninety-degree angle from where I stood. Then I could put my hands "into" the image when I wanted to produce sound similar to hitting a piano key.

This effectively functioned as a manual action segmentation. The experimentation with Soniperforma sparked an interest in exploring how more advanced motion tracking could be used in interactive music systems.

Kinectofon

The ideas from Soniperforma were further developed when I got hold of a Microsoft Kinect "sensor." They call it a sensor, but this device, initially developed for gaming, has two cameras: one regular video camera and one depth-sensitive camera. The latter captures the distance to the objects in the image. This device became the central part of a new prototype instrument, the Kinectofon (Jensenius 2013b). As shown in the signal flow in figure 12.4, Kinectofon uses the same sonomotiongram technique as Soniperforma. However, the Kinectofon allows for filtering the input image based on distance. This is done by combining the image from the depth-sensitive camera with a motion video from the regular camera. The result is a "foreground motion image" based on the camera's desired distance.

It is possible to create a foreground motion image with a regular video stream. However, that requires recording the background first to produce a reliable foreground image. It also helps to have a plain background, such as a single-colored wall. The advantage of a depth camera is that the "noisiness" of the background does not matter. It will effectively be filtered out when setting the depth threshold at the correct level. As I discovered, the depth functionality also allows for creating a performance "plane." I can set the depth threshold half a meter in front of me when performing. Then I can stand in front of the camera and move my hands (and other body parts) "into" the image to create sound. This is an entirely different approach to performing in the air than with Soniperforma. "Touching" the plane almost feels like performing with a physical instrument. The invisible plane is fixed so you can touch the same spot and get the same sonic result. Many people that have tried the instrument have commented about this tactile illusion. The sonic interaction evokes mental imagery of physical touch.

SoundSaber

The air performance interfaces presented so far have all been touchless; the user performs with open hands in the air. An example of an object-based air instrument is the SoundSaber developed by my colleague Kristian

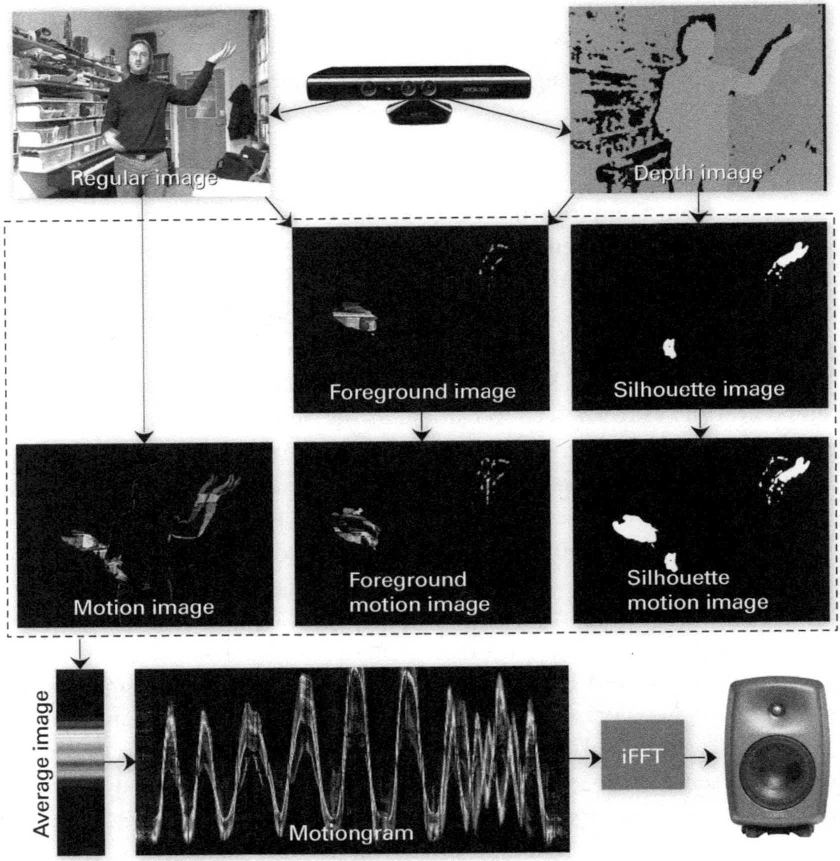

Figure 12.4
An overview of the Kinectofon signal flow. The two input images (regular and depth image) can be processed in various ways before they are passed to the average image module that creates the motiongram. Then the motiongram is converted to sound through the sonomotiongram technique.

Nymoen (Nymoen et al. 2011). This interface is built around a meter-long handheld rod that the performer moves in the air (see figure 12.5). The SoundSaber was initially developed to explore our new infrared motion capture system in the fourMs Lab at the University of Oslo. Later, it was developed into a barebone version using an accelerometer-based system. Common to both implementations was exploring how a seemingly simple sound synthesizer may come to life through high-quality sensing. Many commercial music technology products are opposite: "expensive" sound

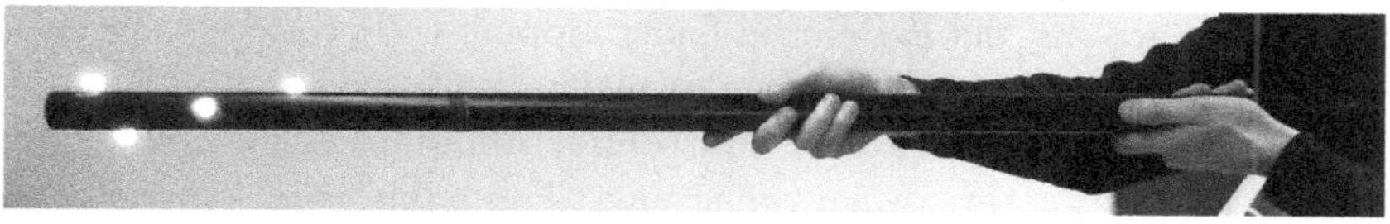

Figure 12.5
The SoundSaber rod has reflective markers attached in a specific constellation so that it is possible to track both its position and orientation. *Source*: Ståle A. Skogstad.

engines with poor control possibilities. SoundSaber exploits the high spatiotemporal resolution and low latency of the motion capture system, allowing for shortening the perceived action–sound separation.

The mappings in SoundSaber were inspired by the many-to-many mappings found in acoustic instruments. One could imagine that tracking only one point in space—the tip of the rod—would lead to a somewhat limited interaction. However, SoundSaber has a clever solution to retrieve as much information as possible from the motion capture data. Attaching multiple markers on the rod makes it possible to define a "rigid object" that can be tracked with both three-dimensional position (X, Y, Z) and three-dimensional rotation (pitch, yaw, roll). This six-dimensional data stream can be used to calculate the absolute and relative velocity and acceleration in both polar and cartesian coordinates.

The SoundSaber's sound engine is based on a sequence of short clicks sent through a chain of delay lines, a ring modulator, another delay line, and a bandpass filter. The first action–sound mapping mimics the "energy transfer" of a sound-producing action by controlling the sound's amplitude with the absolute velocity of the rod. The rod's absolute horizontal position is used to control the sound's spatial location, utilizing the multichannel speaker system in the lab. Thus the user can "move" the sound around in the space. The absolute vertical position is used to control pitch information, with lower frequencies closer to the floor. The combination of absolute and relative mappings means that the user can produce similar sounds independent of the direction faced, similar to what one would expect from a handheld acoustic instrument.

Everyone that has tried SoundSaber comments on its highly expressive nature. I think there are three reasons why it works so well. The first is the high spatiotemporal resolution. We usually run the motion tracking at 200 Hz and with millimeter-level precision. This makes for both immediate and intimate interaction. The mappings are made so that both large and small actions affect the sound. Most users move the rod with large actions, but even rapid finger tapping will produce some audible sound. This creates the feeling of playing an instrument with a large dynamic span, both in action and sound. Another reason SoundSaber works well is its many-to-many mappings. This mimics the complex mapping found in many acoustic instruments. Finally, the rod provides users with the tactile feel of holding onto a metal surface. Its size, shape, and mass give users a haptic experience based on the inertia when moving it around.

The most significant limitation of SoundSaber is that it is bound to our lab. In theory, the setup can be moved elsewhere or be set up in a similar lab elsewhere. However, to test the concept in a more portable format, we recreated the instrument using a Nintendo Wiimote. This controller is based on inertial sensing and cannot track position directly. However, the orientation can be estimated quite precisely, and the accelerometer data can be used to calculate the "energy" used in the instrument. It turned out that only some of the original SoundSaber mappings could be "ported" to the new setup. For example, even though we could extract the angle between the floor and the controller, the orientation data turned out too noisy to be helpful. It was also tricky to recreate the energy transfer function reliably. We, therefore, had to revert to using one of the push buttons on the Wiimote to turn the sound on and off. The reimplemented SoundSaber is fun to play, but it deviates substantially from the original. The fact that it relies on a trigger button also means that it is no longer a "true" air instrument.

Dance Jockey

Dance Jockey is an instrument based on full-body motion tracking using an inertial sensor system (see figure 12.6). It is in many ways the complete opposite of SoundSaber. While SoundSaber is based on multilayered continuous mapping from motion to sound, Dance Jockey uses mappings based on posture recognition. Dance Jockey was developed at the University of Oslo by Yago de Quay and Ståle Skogstad with the aim of catering for interactive dance performances (de Quay et al. 2011; Skogstad et al. 2011). It was from the onset targeted at performances in real-world venues, so we decided to use a commercially available portable inertial motion capture suit (Xsens MVN Biomech). The suit has seventeen inertial sensors attached to the body. Each sensor consists of a 3D accelerometer, a gyroscope, and a magnetometer. These are used to calculate both the three-dimensional position and rotation of a total of twenty-two body

Figure 12.6
A Dance Jockey performance by Yago de Quay in Oslo, November 2010. The performer wears the Xsens MVN motion tracking suit containing seventeen inertial measurement units connected to a wireless transmitter.

joints. This means that the data stream consists of 132 data points coming in at 120 frames per second. The challenge was to figure out how to meaningfully map this high-dimensional data set to sound and music features.

As the name suggests, Dance Jockey was inspired by a "disc jockey" approach to musicking. Instead of creating new sounds from scratch, the performer would trigger samples and add sound effects. This was achieved through a combination of continuous and discrete controls. Since the Xsens suit uses inertial tracking, we decided early on to work with relative position information. That means looking at how much and in which direction motion happens rather than its exact location in space. The system can provide absolute position information, but we experienced a considerable positional drift in our laboratory tests (Skogstad et al. 2011). To ensure stable operation, we instead decided to work with relative motion information.

A benefit of working with full-body motion tracking is that one can calculate relationships between different body joints. In Dance Jockey, we extracted three relative position features describing body limbs and their relationships: the hand distance, elbow angle, and the body's level of contraction or expansion. The latter was implemented as the distance between five body extremities: head, hands, and feet. We also implemented "trigger areas" in the space around the user. These were used to play on virtual "wind chimes" that could be reached by stretching the arms and "touching" the relevant areas around the standing position. The primary shifts in performances were triggered by a hand "clap" feature. The performer could start a new part of the performance by moving his hands closer than 20 cm from each other. This and other static features were calculated using a "posture classifier" based on the angles between different body joints. These angles are relative to the performer's body and are easily reproducible once appropriately trained. Finally, hand speed was used to control the sound level, creating a relationship between the quantity of motion and sound level.

We found that a well-trained posture recognition system is more reliable than action recognition. Discarding temporal information and focusing only on spatial relationships between body parts efficiently reduces the data stream. However, the downside to working with posture recognition systems is that they favor trigger-based mappings. This works well for systems based on sample playback, but it does not allow for continuous control. To draw an analogy to acoustic instruments, one could say that a posture-based instrument is like a piano, while an action-based instrument is like a violin.

The Dance Jockey system was successfully used in several performances. Most performances were based on using an external PA system for the sound playback. We also experimented with the performer carrying the processing computer in a backpack and holding mobile speakers in his hands. Unlike the previously mentioned air instruments, Dance Jockey was more of a music maker than a sound maker. Entire compositions were programmed into the instrument. This made for engaging musical experiences, but the system could not be used for anything else.

Sverm Micromotion Instruments

After researching music-related body motion for a decade, I realized that all my studies and instruments were based on relatively large-scale body motion. I, therefore, decided to shift attention and delve into the world of human "micromotion." This was after I met dancer–choreographer Kari Anne Vadstensvik Bjerkestrand, a tai chi practitioner and slow-motion enthusiast. Together, we began a practice of standing still together. We ran sessions where we stood still on the floor for ten minutes at a time in the motion capture lab. We measured our swaying and carefully wrote down notes about the sessions to compare the qualitative and quantitative data. As can be imagined, it is impossible to stand absolutely still. And when one attempts to stand still for a while, the body's micromotion comes to one's attention. This artistic exploration later turned into a scientific research project in which we have found that people's heads sway at an average rate of up to around 10 mm/s (Jensenius et al. 2017; González Sánchez et al. 2018, 2019; Zelechowska et al. 2020). The exact quantity of motion is not the most important; the main point is that micromotion is significantly smaller than what is normally considered in interactive music systems.

The initial standstill study developed into the artistic research project "Sverm" in which we worked with the micromotion and microsound found when approaching standstill. Here we developed a "spatiotemporal matrix" as a tool for guiding us in rehearsal (Jensenius 2017b). This matrix consists of three spatial and three temporal levels: micro, meso, and macro. Normal actions can be considered as the meso–meso level: in the range of approximately 1 to 100 centimeters and 0.5 to 10 seconds. It is possible to think of a micro–micro action as an action in microspace (less than 1 centimer) and microtime (shorter than 0.5 milliseconds). Anything above 100 centimers and 10 seconds would be considered a macro–macro action. In rehearsals, we tried to perform combinations of micro-, meso-, and macroactions. For example, we could perform a series of micro–micro actions in 10-second intervals for 5 minutes. Then we could switch to performing a micro–macro action over ten minutes. As can be imagined, these extreme cases were the most difficult to master, but they were also the most interesting to work with (Jensenius and Bjerkestrand 2012).

After training up our sensitivity to performing with microactions over several months, we developed what would become a series of instrumental concepts based on what I have called "sonic microinteraction" (Jensenius 2017b). One could argue that many acoustic instruments are built around microinteraction, such as the minute actions found in wind performers' mouths or the fingering of string players. Some electro-acoustic instruments also exploit such detailed control of sound, such as the multidimensional controllers presented in chapter 10. However, there are few examples of air performance instruments exploiting microinteraction. The limitation is not on the sensing side. Nowadays, both inertial and optical motion capture systems are capable of detecting micromotion, something we have documented in various technology experiments (Skogstad et al.

Figure 12.7
Rehearsing with the Waving Sines instrument for one of the performances of the "Sverm" project. Each performer inversely controls a sine tone based on the tracked position of a reflective marker on the head.

2011; Jensenius et al. 2012; Nymoen et al. 2012; Bishop and Jensenius 2020). In my experience, the limitations are mainly on the conceptual side.

Waving Sines was my first microinteraction instrument (Jensenius 2017b). It was conceived after some initial tests of sonification of the continuous motion data stream from the motion tracking system. I decided to map the vertical position of a motion capture marker directly to the frequency of a sine tone generator. This is probably the simplest motion–sound mapping, yet it turned out to be remarkably interesting. Sine tones are usually perceived as static, but when controlled by a high-speed motion tracking system, the sound came "alive." I continued the testing by mapping the quantity of motion to the sound level. This made it possible to make sound only when moving. At some point, I did something wrong in the programming and ended up with a reverse mapping; sound would only appear when there was no motion. If one moved, there would be no sound. This is counterintuitive, yet we all felt that it was inspiring to work with. In Waving Sines, we utilized this "inverse control" paradigm in performance (see figure 12.7). The performers each wore a reflective marker on their heads. Since the marker's vertical position was mapped to the frequency of a sine tone, a taller person would produce a higher pitch than a lower one. The horizontal position controlled the spatial location of sound when played back over a multichannel sound system. The twist is that sound is generated by standing still; motion leads to silence.

Performing with an instrument built around inverse sonic microinteraction is more complicated than it sounds. Such a mapping is counterintuitive by design. That is also what makes it fascinating. Once you master producing sound by standing still, you want to try modifying the pitch. However, this requires moving (a little). However, as soon as you move, the sound disappears. As we rehearsed with the instrument, the performers chose different strategies. Some just gave up on controlling it continuously and decided to play by standing still for a while, then moving to a new posture to get another pitch. Others managed to move very slowly while maintaining a continuous sound. This works if your motion is small and slow enough. However, due to the high accuracy and precision of the motion capture system, such micromotion will lead to small fluctuations in both amplitude and frequency of the sound. That is what makes the sonic result of the instrument interesting. Listening to a static sine tone can be annoying. However, listening to multiple sine tones that continuously change in both amplitude and frequency makes the sound come alive. The sounds are shaped by the swaying and breathing of the performers.

Waving Sines can be performed alone, but it was developed as a collaborative instrument. I particularly enjoy listening to how the different sine tones blend when multiple people perform. If you have people of the same height—or if they adjust their heads to the same vertical position—one may get beat frequencies. That is, one may hear a pulsation based on the difference between the individual frequencies. Once, I had a group of students who spent a good half hour aligning themselves to create such complex beat patterns. Another group explored playing harmonies together by positioning their heads accordingly. Of course, all such explorations are made more challenging because sound disappears once you move.

Muscle Performance

Waving Sines and the other microinteraction-based instruments we developed in the "Sverm" project were all based on camera-based motion tracking. To move outside of the lab, I got curious about testing other sensor types and sensing modalities. Then muscle sensing quickly comes to mind. Muscle activity can be measured using electromyography (EMG), one of the most commonly used biosignals for musical applications (Aly et al. 2021). Knapp and Lusted (1990) did pioneering work on EMG-controlled music in the 1990s. Their BioMuse system has been used by several others, of which the work of Tanaka (1993) is perhaps the most well known. There are also examples of using muscular sound in performance, such as in the works by Donnarumma (2016). His Xth Sense system can be seen as a hybrid electro-acoustic instrument since it amplifies and modifies acoustic sound generated in the body.

Specialized muscle-sensing equipment has been available for decades. Still, it was the advent of the commercially available Myo armband controller in 2014 that sparked a broader interest in using muscular control in interactive music systems. Not only is the Myo armband easy to use with its eight muscle sensors, accelerometers, wireless connection, and rechargeable battery, but the data is also of sufficiently high quality to be used for musical applications (Nymoen et al. 2015). We also verified that it could sense microlevel activity (Jensenius et al. 2017). Unfortunately, the armbands went out of production after some years, but we still have a collection of devices that we continue to explore at the University of Oslo.

One of the instruments we have built around a Myo armband is the MicroMyo instrument. This is a complete instrument containing a small Bela-based microcomputer (McPherson and Zappi 2015), a battery pack, and a portable speaker (Martin et al. 2018). This portable design makes it easy to perform in any location. In the piece *Stillness Under Tension*, we explored the MicroMyo instrument in a collaborative performance setting (see figure 12.8). Here the mappings were based on the same inverse control principle as Waving Sines. The performers need to stand still for sound to appear. When the performer moves, the sound fades out quickly.

The Myo armband provides relative motion information through the inertial sensor. Here we found the data from the embedded gyroscope to be most reliable. So we could use arm rotation to control the base frequency of the sound engine. Holding the arm down will result in a lower pitch. Data from the eight muscle sensors were mapped to frequencies of individual partials in an additive synthesizer. This made it possible for each performer to control the sound's timbre by flexing their arm muscles. They had to do this carefully since the sound would fade out if they moved too much. Similar to Waving Sines, the performers had a constant "battle" between trying to move and not move at the same time.

Even though MicroMyo and Waving Sines are based on the same conceptual idea, their implementation is entirely different. In Waving Sines,

Figure 12.8
The premiere of *Stillness Under Tension* in Oslo in November 2017. Each performer wears a Myo armband on their arm, capturing motion, rotation, and muscle tension in the arm. *Source*: Simen Kjellin.

the performers wear a single, lightweight marker on top of the head. This encourages standing upright and focusing on swaying back and forth or moving slowly up and down. The sensor armbands used in MicroMyo focus the attention of both performers and perceivers on the arm. Thus the two instruments afford completely different musical interactions.

Waving Sines is based on a multicamera motion capture system, a multichannel sound system, and two PCs. In theory, it is possible to move the setup. In practice, it is only used in our lab. MicroMyo, on the other hand, is a portable instrument. The complete instrument can fit in a small bag, including the handheld speaker. The result is that I regularly both demonstrate and perform with MicroMyo in various settings, while I only occasionally start up Waving Sines for a demonstration in the lab. The ideas from MicroMyo have also developed into other similar setups, such as in the music–dance piece Vrengt (Erdem et al. 2020).

I want to dwell on the fact that MicroMyo is running on an embedded system. The prototype of the instrument was developed on a laptop (Jensenius et al. 2017). This worked, but we found it unreliable in performance, primarily because of Bluetooth connectivity problems. The setup did not scale well either since one laptop was needed per armband. A six-person performance required a quite large technical rig, with the added complexity of simultaneously getting six computers up and running. There was always some issue: a software update notification, the failure to load a package, or a Bluetooth connectivity problem. These are common challenges for computer-based setups and are usually easy to solve. Still, I find such issues problematic in performance contexts with limited setup time and where the systems need to be left alone for some time between setup and performance. Then working with embedded systems is much more comfortable. After reimplementing MicroMyo using small Bela computers (Martin et al. 2018), the whole setup consisted only of the armband, the microcomputer, a battery, and a speaker. This meant that each performer could carry their instrument and be in complete control of its operation. The importance of using single-purpose microcomputers in musical instruments should not be underestimated (Berdahl and Ju 2011). Having an embedded system has made the instrument more user-friendly. One just puts on the armband, turns on the microcomputer and speaker's power, and starts producing sound.

My excitement for embedded systems coincides with a general interest in more "hardware"-based devices. I am writing "hardware" here because while MicroMyo and other similar instruments may be considered standalone devices, they contain a computer. The main difference is that it is a dedicated and stripped-down computer that is only supposed to do one thing. It is interesting to witness the "cyclic" nature of technological development. At the beginning of the digital era, all systems were single-purpose devices. As more powerful computers appeared, people would interact with these servers through "thin" clients. These thin clients were overthrown by the personal computers in the 1980s, laptops at the turn of the millennium, and tablets and mobile phones since the 2010s. Today, we see an increased interest in server-based solutions and embedded systems again. One reason for this is that larger and more powerful does not necessarily mean faster. McPherson et al. (2016) have shown that an embedded platform can have a lower action–sound latency than a laptop-based system. Making complete standalone instruments most likely also ensures a longer lifetime for those instruments.

Self-Playing Guitars

The last series of air instruments I will present in this chapter is based on a concept we call the "Self-playing guitars." We have developed this platform over the last few years at the University of Oslo, resulting in several different installations and performances. The first of these, the installation *Sverm–Resonans*, consisted of six actively augmented acoustic guitars (see figure 12.9). Each of these could be considered a self-contained

Figure 12.9
A performance with the installation *Sverm–Resonans*. Each guitar is an independent instrument controlled by the microinteractions of a standing-still human.

electro-acoustic instrument, and the whole installation could probably best be described as a large meta-instrument.

The Self-playing guitars are not really "touchless." In fact, we often encourage touching and holding them. However, the control of the guitars is touchless, and conceptually they have been developed based on the ideas of standstill and inverse microinteraction that came out of the *Sverm* project. The main difference between the Self-playing guitars and the other instruments I have described earlier in this chapter is the focus on the electro-*acoustic* sound played through the body of the guitars. There has been a growing interest in the augmentation of acoustic instruments in recent years, ranging from augmentations aimed at helping people to learn an instrument (Baldwin et al. 2016; Heller et al. 2017) to tools for expert performers (Kimura et al. 2012; Lähdeoja 2015; Reboursière et al. 2010). One reason for this trend may be to exploit the richness and nonlinearities of acoustic sound generation while simultaneously utilizing the power of digital sensing and control.

The artistic goal of the installation *Sverm–Resonans* was to invite people to connect with their breathing and give them time and space to relax. This was something we had seen that people responded positively to in the *Sverm* performances. Now we wanted to see if we could generate some of the same effects in a gallery space. The installation was conceived as a meeting point between a living body interacting with an electronic sound system played through a vibrating acoustic instrument. As such, *Sverm–Resonans* explored the meeting points between the tactile and the kinesthetic, the body and the mind, and between motion and sound. Each guitar produced a quiet low-frequency drone while hanging. This created a sonic ambiance in the gallery space. They were mounted in thin fish lines, which made them rotate slowly. The result was a stable yet continuous sound in the space.

The interaction with the guitars was based on the same inverse control idea explored in some of the previous instruments. The guitars had an infrared sensor attached to the neck, which would detect a person standing in front. After a few seconds of standstill, the guitars would gradually produce sound. This was a slow breathing-like sound aimed at helping people to focus on their respiration cycle. The pitch of the breathing sound would slightly change based on a person's distance to the guitar. If moving beyond a specified threshold, the breathing-like sound would disappear. Then one had to stand still again and wait for the sound to reappear. Working on the prototype, we discovered the tactile and haptic experience of holding the vibrating guitars. For the installation, we therefore encouraged people to hold the guitars. Many did, but the interaction also worked if someone just stood still without holding.

Each of the guitars is an independent and self-contained instrument, equipped with a microcomputer, various sensors, an actuator, and a battery pack (González Sánchez et al. 2018). There are no external speakers; all sound is generated by the actuators mounted on the back of the guitars. On the sensor side, one infrared distance sensor measures the proximity of people, an inertial measurement unit picks up motion and rotation, and a microphone "listens" to sounds in the space. Together, this provides a flexible and powerful electro-acoustic instrument. The guitars can easily be reprogrammed using various music programming languages (PureData, Csound, SuperCollider). One thing that has become clear is the need to fine-tune the sound engines to the instrument. Standard speakers have a relatively flat sonic response, while the guitars have a strong sonic identity. Therefore, it is essential to develop sound engines and create mappings between sensors and the sound features that match the guitars' sonic characteristics. This has been a useful reminder about the importance of considering the whole chain from action to sound when designing instruments.

Over the years, we have explored many different sound designs with the guitars. The *Sverm–Resonans* installation was meditative. In *Sverm–Puls*, we explored a rhythmic sound design based on entrainment principles. Entrainment is the process with which individual rhythmic patterns interact with each other (Clayton et al. 2020). Our entrainment implementation was based on the firefly-inspired design suggested by Nymoen et al. (2014). When left alone, the guitars entrain to a common pulse based on time shifting their pulsations to a common pulse. This is done by "listening" to the environment using built-in microphones. The idea is that other sounds "distract" the entrainment process, thereby creating a more complex soundscape. The infrared sensor controls the frequencies of the pulsating tones. So it is possible to play with the guitars by placing oneself in front of them.

It has been exciting to see how people have approached the different installations we have set up with the Self-playing guitars. Many have responded positively about being allowed to touch the instruments. Performers are used to touching sound-producing objects, but perceivers primarily listen to music through their ears. The tactile and haptic elements—in addition to the sonic and visual—enriches the experience. One participant described the installation in this way:

> I enjoyed it. I was sensing what it gave to me and how I could meet it with different parts of my body, because also I am a dancer, and a yoga teacher, so I am physically oriented, which was my starting point. . . . And it was interesting to see that something new developed over time with this relationship with this instrument, which I merely was holding. . . . It was playing itself, or playing me.

The Self-playing guitars are still in active exploration in our lab. What is particularly interesting is how having six similar self-playing guitars allows for explorations into multiperformer setups. In some ways, the concept borrows ideas from the tradition of standardized laptop orchestras and mobile phone ensembles. There, too, multiple performers play on the same type of instrument. Still, the guitars are different. They are more limited in what they can do because of the limited sensing and processing capabilities. They also have a quite distinct sound profile, since all sounds are colored by the guitar bodies. However, this constraint can also be used creatively when designing new action–sound mappings.

From Touchless to Touching

Some people think about air instruments as a "gimmick," something science fiction-like. However, air instruments are real, and they are increasingly used in all sorts of performances. I find air instruments fascinating from an action–sound perspective. Air instruments are by design disembodied. Furthermore, it is impossible to find anything less causal than creating instruments based on inverse control and autonomous processes. Nevertheless, I have found that exploring such noncausal interaction forms is relevant for learning more about musical human–computer interaction. This is also relevant to our understanding of music cognition at large.

Common to all of the instruments presented here is that they use computationally "cheap" sound engines. Almost all are based on simple additive synthesis or noise as the starting point and some basic filters and feedback. More focus has been put into creating faster and more responsive control. As such, these instruments are opposite to many other instruments in which the focus is put on creating advanced sound engines. A system is never better than its weakest part, and if you have poor control, you will struggle to make a high-quality sonic interaction. Many of the instruments mentioned above have shown that you can create a high level of intimacy and immediacy through a high spatiotemporal richness in the control data. Rapid and low-latency control provides the feeling of "touching" the sound. This was evident already when I developed the Kinectofon, in which the depth camera allowed for creating a virtual wall in front of the performer that one could touch.

It may seem like a contradiction to end a chapter on air instruments with a description of the Self-playing guitars. After all, one of the most exciting parts of the guitars is that you can feel the sound vibration in the instrument body. Nonetheless, their interaction is touchless; it is their output that is based on touch. Here the combination of sonic and haptic feedback creates a truly cross-modal experience. The same can be said about SoundSaber. That instrument is also performed by moving in the air, but the inertia of the moving rod is an essential part of the user experience. There is much potential in exploring such multimodal and crossmodal interaction modes in interactive music systems.

Postlude

"What happens if I turn it the other way?" my youngest daughter shouted in excitement as she pointed at the rotary dial on the new radio I had just picked up during our Stockholm vacation. She had just tested the "scratching" function of the Teenage Engineering OB-4, branded as a "magic" radio. The "magic" part of the radio is the implementation of a synthesizer module into the device. A built-in continuous recording mode makes it possible to rewind whatever one listens to and adjust the playback speed. Instead of only listening passively to the radio, my daughters started modifying the sound. They quickly discovered how to create loops of various durations. I was equally excited about the ambiance mode, which makes a "sonic soup" of the incoming audio based on a long delay line. "What do you call such a thing?" my wife asked, looking at me to get a straightforward answer. "Well, it is kind of . . . kind of . . . a musicking technology," I replied, trying to come up with a better term than "magic" to describe its function. After all, there is nothing magical about the device. Instead, it is an excellent meeting point between creative art and sound engineering. "This device is in many ways a real-world example of the conclusion of my new book," I added, trying to explain my thoughts on how new technologies allow for active musical experiences.

Musicking Technology

As I have argued throughout this book, new technologies change the way we perform and perceive music. New instruments are developed every year. Most of them are never commercialized, and many that reach the market fail to make an impact. There is also much experimentation when it comes to music technologies targeted at music listening. Both types of technologies—targeted at performers or perceivers—challenge the traditional idea of what an instrument should be and how it should function.

They expand the action–sound palettes available, the spatiotemporal distance between instruments and performers, and the resultant musical output. Not least do they blur the lines between different actors. Today, many composers build instruments, record sounds, and perform themselves. Similarly, performers may spend more time programming sounds and musical structures before a performance than they do on stage. In parallel, perceivers get access to musical apps that let them create music on their own or modify prerecorded music. In many ways, the processes of making and experiencing music are changing.

New musicking technologies have a different scope and intention than traditional music technologies. Today's new instruments are not only sound makers; they are often also music makers with much embedded musical knowledge. In some cases, the instrument may contain a more or less fixed piece of music or artificial intelligence that can create music on the fly. Other times, a musicking technology may allow for full musical exploration or tweaking of the sonic output in the case of the magic radio. The development of such devices is escalating. However, as I have argued in earlier chapters, the shift from sound makers to music makers began centuries ago. The church organ made it possible to change the sound's timbre and play harmonic tones. Music boxes embedded whole compositions that could be played back at different speeds. The shift from sound makers to music makers happened long before electricity came into play.

"Children should play the piano, not on a mobile phone," a parent once told me when I talked about my research. Fortunately, I do not think that new musicking technologies need to overthrow other musical activities. My daughters play with all the new electro-acoustic instruments I bring home. They also enjoy playing trombone, saxophone, flute, and piano. All these instruments allow for different sonic and musical experiences, and they can perfectly well coexist. People still play acoustic instruments even though electro-acoustic instruments have been available for decades. I am not afraid that devices like the magic radio will lead to a decline in piano playing. However, new musicking technologies may allow for more active and embodied musical experiences. They may open for more playful musical engagement than only passively listening to the radio.

How should we think about musical instruments in the future? Should all the emerging musicking technologies be regarded as instruments? In many ways, trying to define an instrument may cause more harm than good. One could be conservative and say that an instrument should allow for real-time musicianship with a low action–sound separation. That

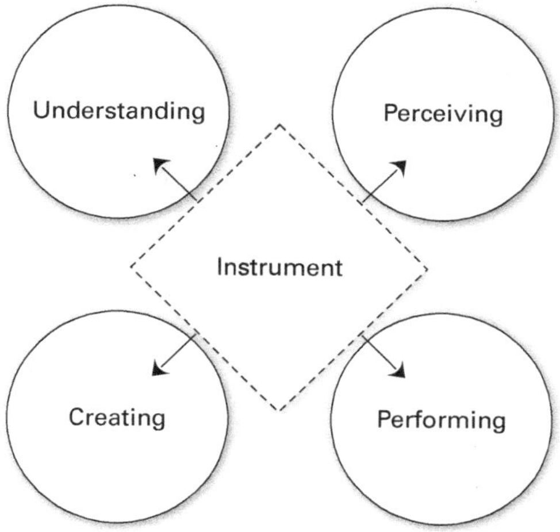

Figure 13.1
New instruments challenge the traditional roles of the musicking quadrant.

would rule out most new devices. Instead, one could be liberal and argue that any musicking technology should be considered an instrument. Then one risks "watering out" the instrument concept to such an extent that nobody understands its purpose. The musicking quadrant can help in structuring our thoughts on the matter. As shown in figure 13.1, new instruments may be thought of as bridging over between different musical functions, roles, and actors. What is clear is that a new "instrument" definition will cover more than only sound makers. In my thinking, instruments need to allow for active and embodied musical interaction within a certain level of action–sound separation and spatiotemporal distance. There needs to be a musical *inter*action. Thus defining instruments based on their techno-cognitive properties is more valuable than looking only at their technical construction.

Differences between Couplings and Mappings

One of my core arguments has been that there are techno-cognitive differences between action–sound couplings and mappings. I introduced this distinction in my PhD dissertation (Jensenius 2007), although the argument was not yet fully developed at the time. My focus on action–sound couplings was inspired by recent trends in embodied music cognition. *Ways of Listening* by Eric Clarke (2005) had just come out, and I had read a draft manuscript of *Embodied Music Cognition and Mediation Technology* by Marc Leman (2008). My clear-cut separation between the couplings of acoustic instruments and the mappings of electro-acoustic instruments was based on speculations about how our brains work. There is a growing body of neuroscientific research into relationships between action and sound in general. However, less is known about possible cognitive differences between action–sound couplings and mappings.

It should be possible to investigate some of these differences scientifically. For example, is there a difference in the way we perceive an acoustic and a digital piano? Such instruments may look and sound more or less the same. My hybrid piano feels almost the same as playing an acoustic grand piano. Still, I believe there are differences. For example, an acoustic instrument will almost always be slightly out of tune. However, the most significant conceptual difference is that one instrument runs on electricity. If the power is off, there is no sound. When I hit the key on an acoustic piano, it will always sound with a piano-like tone. When I press the key on an electro-acoustic piano, I never know what will happen. It may not sound if it is not turned on, the volume control is down, or the power is out. If it sounds, there is no way that I could have predicted its sonic qualities. This makes me approach acoustic and electro-acoustic instruments differently. The same is the case when others perform. My expectations are shaped by the instrument I see. If someone picks up the violin, I know more or less what to expect. If they sit with a laptop, I have no idea. Is that a problem? Not necessarily. Several of my strongest musical experiences have been with electro-acoustic instruments. But I have also experienced many unsuccessful performances with such instruments.

My attempt at separating acoustic and electro-acoustic instruments is not based on a wish to say that one is better than the other. Instead, I have created a theoretical framework that can help us understand more about musicking technologies, regardless of whether we are talking about a violin or a new electro-acoustic instrument. Hopefully, this can also help in discussing the qualities of various instruments. One could argue that this is primarily of theoretical interest. However, such questions become real when I sit on student entry committees in my department. It is not uncommon to compare one student playing classical music on a clarinet to another presenting a live electronics piece on a laptop. It is impossible to make informed judgments about such performances if we are not equipped with a vocabulary to describe the instruments in question and the different types of musicking that they afford. This book does not provide a solution for making such judgments. However, it can hopefully help in understanding more about different types of musicking.

Diversity and Accessibility

Several topics have been left out or only marginally discussed in this book. My main story has focused on the individual's experience with instruments. In the future, it would be exciting to combine such a techno-cognitive perspective with "sociotechnical" theories, such as the one described by Bijker (1997). A broader sociocultural perspective could have opened for discussing diversity more broadly. The male dominance of the music technology research field is well documented. Xambó (2018) showed that only around 14 percent of authors at the annual NIME conferences are women. This corresponds with the low female numbers found by Born and Devine (2015) in a study of music technology education in the United Kingdom. There are also few women in the music technology performance scene, as documented by Gadir (2017) in studies of female DJs. Based on interviews with women experts in the field, Jawad and Xambó (2020) argue that the term "music technology" may in itself be problematic.

There are also other diversity issues in the music technology community. Frid (2019) has described the need for more focus on accessibility and inclusivity in musical interaction design. Morreale (2021) has critized the neocolonial tendencies of current music technologies. Finally, there has also been a widened understanding of the challenges with the "white racial frame" that limits the topics, discourse, and job opportunities in music research more generally (Ewell 2020). My approach is shaped through the "frame" of a white northern-European male. Until recently, I did not explicitly focus on diversity or accessibility in my research. However, my growing awareness of these issues has influenced some of my choices when developing the theories presented in this book. I hope to make a difference by pushing the musicking technology concept and experimenting with building prototypes of such devices. I also hope that it helps (a little) to acknowledge the diversity challenges of the various fields I am involved in.

Sustainability

Sustainability is another topic that is quickly emerging based on the environmental crisis we are currently experiencing. I am generally optimistic about the possibilities of new technologies. For example, I believe that developing better solutions for telematic performance may help decrease the need to travel in the future. However, I am increasingly concerned about today's technological consumption, which is unsustainable. There has been a growing focus on sustainability in general human-computer interaction (Chapman 2015; Hazas and Nathan 2017) but less so in the music (technology) research community. Devine (2019) talks about "petrocapitalism" in his study of the negative environmental impact of the music industry. Even though there are industries with larger impacts, that is no excuse for not making a change.

Mankoff et al. (2007) have argued that technology designers can contribute by keeping sustainability in mind when developing new products. This can be done from two perspectives:

- Sustainability through design
- Sustainability in design

King and Chang (2016, 204) exemplify sustainability through design with an innovative water dispenser. The dispenser has a built-in counter that lets the user know how many plastic bottles have been saved by using the machine. It is a small design inclusion but has a "game-like" effect when giving users positive feedback. I have yet to come across a good example of a musical instrument focusing on sustainability through design. There are also, unfortunately, few examples of sustainability in design in the music technology community (Masu et al. 2021). One example that comes to mind is the Fingerphone by Freed (2012). This project was based on recreating a popular instrument from the 1970s, the Stylophone, with a more sustainable technological implementation. His approach was based on reducing the number of components in the device. He also reused components from other devices, such as the built-in battery. Finally, he built a large part of the device with recyclable paper, using principles by Koehly (2011). Through these simple measures, Freed showed that it is possible to make a complete musical instrument with a relatively small environmental footprint.

I have several architect friends, and I hear many of them talk about the future of their profession being *re*construction rather than construction. We should probably focus more on reconstruction also in the music technology community. Here we can learn from hardware hackers, experts in reusing and repurposing technologies for musical purposes (Collins 2020). We can also build on the exisiting tradition of building and maintaining traditional musical instruments. Instruments have for centuries been

Table 13.1
A framework for reflecting on the environmental impact of a musical instrument (Masu et al. 2021).

		Research stage			
		Making	Testing	Using	Disposing
Resources	Materials				
	Consumables				
	Storage				
	Processing				
	Transport				

built to last. It is only in recent years that consumer electronics have been designed to break. Thus creating with quality in mind is essential. "Designing for disassembly" is another (Crowther 2005). That may help to fix broke instruments but also facilitate repurposing.

How can one get started with a more sustainable music technology practice? Table 13.1 shows a framework we proposed for considering the environmental cost of activities related to designing, building, and using instruments (Masu et al. 2021). The matrix is not exhaustive but is a tool for considering one's environmental footprint. The columns represent the research stage. The first is the making of an instrument, which covers design, construction, prototyping, and coding. Next is the testing phase, including performances made as part of an iterative design phase. The using phase covers regular rehearsals and performances. Finally, disposing covers the end-of-life stages of an instrument. Five different topics can be considered for each step. The materials are the physical parts necessary to build a device. The consumables are physical parts that need to be regularly replaced, such as the reeds of wind instruments and markers of motion capture–based instruments. Storage covers the environmental cost of both physical and digital storage. Processing is the cost of computing, which is mainly relevant for instruments built with CPU-intensive AI components. Finally, transport covers all sorts of transportation related to the different phases, ranging from the transport of instrument parts to travel as part of performance activities.

We introduced the framework in a workshop at the NIME conference in 2021. The participants tried to fill out the matrix for one of their instruments. Many found it challenging to fill all the cells. Still, it was perceived as a valuable tool to help structure one's thoughts. Several argued that transportation—either their own or related to parts and equipments—is what drives their environmental cost the most. Some said they had not thought about the environmental cost of adding batteries to their instruments. Others started reflecting on the use of intensive machine learning for their sound engines. Personally, I started to think about the environmental footprint of keeping an extensive collection of instruments in physical storage. That storage space needs heating, and it could probably have been used for other things. It made me realize that I need to reduce the number of instruments I keep in storage. Some should be put to use again; others should be passed on to others or be repurposed.

Three Design Principles

When I began writing this book, I thought about ending with a set of concrete guidelines for the design of musical instruments. I have over the years been inspired by the thirteen principles for designing computer music controllers by Cook (2001) and the ten laws of simplicity by Maeda (2006). Over time, the project became more theoretical and my principles gradually faded. What is left are three core concepts that I believe are important to consider: sound, intimacy, and immediacy.

First, we should care more about the sound of instruments. This may seem obvious, but we have seen a gradual separation between action and sound in new instrument designs over the last decades. Paradoxically, we also see that sound compression, streaming services, and miniaturization of electronics lead to a deterioration of the perceived sound quality. My action–sound theory is a tool to help discuss and reflect on the whole chain from action to sound.

Second, I believe that the feeling of intimacy with an instrument is related to its action–sound separation. One of the joys of playing an acoustic instrument is the tactile and haptic experience of touching the vibrating object(s). Much work remains before we reach a similar level of intimacy in electro-acoustic instrument designs. Building in speaker elements and actuators in the instrument body is a good starting point. Developing various types of hybrid instruments is another. Also, as demonstrated in some of my air instruments, it is possible to create intimate interaction experiences through clever multimodal designs.

Third, the feeling of intimacy is related to the sense of immediacy. Playing in time and in real time helps with experiencing that something happens "here" and "now." Interactive systems based on high sampling rates and low latencies generally feel more immediate than others. As such, one might believe that telematic systems are by definition out-ruled. However, new communication technologies may increase the physical distance yet retain a low perceived spatiotemporal distance between action and sound. Network technologies are already changing the musical ecosystem. It will be exciting to continue exploring such musicking technologies in the future.

Bibliography

Aho, M. (2016). *The Tangible in Music: The Tactile Learning of a Musical Instrument*. SEMPRE Studies in the Psychology of Music. Abingdon, UK: Routledge.

Ahrens, C., and I. Zedlacher (1996). Technological Innovations in Nineteenth-Century Instrument Making and Their Consequences. *The Musical Quarterly 80*(2), 332–340.

Alarcón Díaz, X., and A. R. Jensenius (2019). "Ellos no están entendiendo nada" ["They are not understanding anything"]: Embodied remembering as complex narrative in a Telematic Sonic Improvisation. In *Proceedings of RE:SOUND*. BCS Learning & Development.

Alarcón Díaz, X., P. Boddie, C. Erdem, E. Aandahl, E. S. Andersen, E. Dahl, M. Lesteberg, and A. R. Jensenius (2019). Sensing Place and Presence in an INTIMAL Long-Distance Improvisation. *Journal of Network Music and Arts 1*(1), 1–22.

Aly, L., H. Silva, G. Bernardes, and R. Penha (2021). Appropriating Biosensors as Embodied Control Structures in Interactive Music Systems. *Human Technology 17*(1), 45–80.

Ansdell, G. (2014). *How Music Helps in Music Therapy and Everyday Life*. New York: Routledge.

Applebaum, S., and T. Lindsay (1986). *The Art and Science of String Performance*. Van Nuys, CA: Alfred Music.

Askenfelt, A., and E. V. Jansson (1990). From touch to string vibrations. I: Timing in the grand piano action. *The Journal of the Acoustical Society of America 88*(1), 52–63.

Association, T. M. (2020). MIDI 2.0 Specifications. Technical report, The MIDI Association.

Association, T. M. (2018). MIDI Polyphonic Expression Version 1.0. Technical report, The MIDI Association.

Association, T. M. (2020b). Common Rules for MIDI-CI Profiles, Version 1.0. Technical report, The MIDI Association.

Baalman, M. (2022). *Just a Question of Mapping*. Rotterdam: V2_.

Baldwin, A., T. Hammer, E. Pechiulis, P. Williams, D. Overholt, and S. Serafin (2016). Tromba Moderna: A Digitally Augmented Medieval Instrument. In *Proceedings of the International Conference on New Interfaces for Musical Expression*, Volume 16 of *2220-4806*, Brisbane, Australia, pp. 14–19. Queensland Conservatorium Griffith University.

Barbosa, J. (2019). "Direct Manipulation for Musical Interfaces Based on Visual Software: Design and Evaluation." PhD diss., McGill University.

Barlindhaug, G. (2019). "The Kids Want Noise - How Sonic Mediations Change the Aesthetics of Music." PhD diss., UiT The Arctic University of Norway.

Barry, B. R. (1990). *Musical Time: The Sense of Order*. Stuyvesant, NY: Pendragon Press.

Bartók, B. (1976). Mechanical Music. In *Béla Bartók Essays - Selected and Edited by Benjamin Suchoff*. London: Faber & Faber.

Bates (2012). The Social Life of Musical Instruments. *Ethnomusicology 56*(3), 363.

Bell, A. P. (2018). *Dawn of the DAW: The Studio as Musical Instrument*. New York: Oxford University Press.

Berdahl, E., and A. Kontogeorgakopoulos (2013). The FireFader: Simple, Open-Source, and Reconfigurable Haptic Force Feedback for Musicians. *Computer Music Journal 37*(1), 23–34.

Berdahl, E., and W. Ju (2011). Satellite CCRMA: A Musical Interaction and Sound Synthesis Platform. In A. R. Jensenius, A. Tveit, R. I. Godoy, and D. Overholt (Eds.), *Proceedings of the International Conference on New Interfaces for Musical Expression*, Oslo, Norway, pp. 173–178.

Bevilacqua, F., and R. Muller (2005). A gesture follower for performing arts. In *Proceedings of the International Gesture Workshop*.

Bielawski, L. (1979). Instrumentalmusik als Transformation der menschlichen Bewegung. Mensch-Instrument-Musik. *Studia instrumentorum musicae popularis 6*, 27–33.

Bijker, W. E. (1997). *Of Bicycles, Bakelites, and Bulbs: Toward a Theory of Sociotechnical Change*. Cambridge, MA: MIT Press.

Birnbaum, D., R. Fiebrink, J. Malloch, and M. M. Wanderley (2005). Towards a Dimension Space for Musical Artifacts. In *Proceedings of the International Conference on New Interfaces for Musical Expression*, Vancouver, pp. 192–195.

Bishop, G., G. Welch, and B. D. Allen (2001). Tracking: Beyond 15 minutes of thought. In *SIGGRAPH Course 11*, Los Angeles, pp. 6–11.

Bishop, L., and A. R. Jensenius (2020). Reliability of two IR Motion Capture Systems in a Music Performance Setting. In *Proceedings of the Sound and Music Computing Conference*, Torino, Italy, pp. 47–53.

Bishop, L., V. Gonzalez Sanchez, B. Laeng, A. R. Jensenius, and S. Høffding (2021). Move like Everyone is Watching: Social Context Affects Head Motion and Gaze in String Quartet Performance. *Journal of New Music Research 50*(4), 392–412.

Bjørkvold, J.-R. (1992). *The Muse Within*. New York: HarperCollins.

Bjørn, K., and C. Meyer (2018). *Patch & Tweak - Exploring Modular Synthesis*. Copenhagen: Bjooks.

Boethius, A. M. S. (1989). *Fundamentals of Music: Anicius Manlius Severinus Boethius*. New Haven, CT: Yale University Press.

Bongers, B. (2000). Physical Interfaces in the Electronic Arts : Interaction Theory and Interfacing Techniques for Real-time Performance. In M. M. Wanderley and M. Battier (Eds.), *Trends in Gestural Control of Music*, pp. 41–70. Paris: IRCAM – Centre Pompidou.

Borgdorrf, H. (2012). *The Conflict of the Faculties - Perspetives on Artistic Research and Academia*. Leiden, Netherlands: Leiden University Press.

Born, G., and K. Devine (2015). Music Technology, Gender, and Class: Digitization, Educational and Social Change in Britain. *Twentieth-Century Music 12*(02), 135–172.

Boulanger, R. C. (Ed.) (2000). *The Csound Book: Perspectives in Software Synthesis, Sound Design, Signal Processing, and Programming*. Cambridge, MA: MIT Press.

Boulanger, R. C., and V. Lazzarini (Eds.) (2011). *The Audio Programming Book*. Cambridge, MA: MIT Press.

Bovermann, T., A. de Campo, H. Egermann, S.-I. Hardjowirogo, and S. Weinzierl (Eds.) (2017). *Musical Instruments in the 21st Century*. Singapore: Springer Singapore.

Bowers, Q. D. (1972). *Encyclopedia of Automatic Musical Instruments*. Vestal, NY: Vestal Press.

Braasch, J. (2009). The Telematic Music System: Affordances for a New Instrument to Shape the Music of Tomorrow. *Contemporary Music Review 28*(4-5), 421–432.

Brandenburg, K. (1999). MP3 and AAC Explained. In *AES 17th International Conference on High Quality Audio Encoding*.

Brandtsegg, Ø. (2022). Towards a pitch map for vibrotactile feedback instruments. *Echo* Vol. 3.

Brandtsegg, Ø., T. Engum, and B. I. Wærstad (2018). Working methods and instrument design for cross-adaptive sessions. In T. M. Luke Dahl, and Douglas Bowman (Eds.), *Proceedings of the International Conference on New Interfaces for Musical Expression*, Blacksburg, VA, pp. 1–6. Virginia Tech.

Bregman, A. S. (1994). *Auditory Scene Analysis*. Cambridge, MA: MIT Press.

Briot, J.-P., G. Hadjeres, and F.-D. Pachet (2019). *Deep Learning Techniques for Music Generation*. Cham, Switzerland: Springer.

Brøvig-Hanssen, R. (2018). Listening to or Through Technology: Opaque and Transparent Mediation. In S. Bennett and E. Bates (Eds.), *Critical Approaches to the Production of Music and Sound*, pp. 195–210. New York: Bloomsbury Academic.

Brøvig-Hanssen, R., and A. Danielsen (2016). *Digital Signatures: The Impact of Digitization on Popular Music Sound*. Cambridge, MA: MIT Press.

Browman, C. P., and L. Goldstein (1989). Articulatory gestures as phonological units. *Phonology* 6(02), 201–251.

Brown, A. R. (2012). *Computers in Music Education: Amplifying Musicality*. New York: Routledge.

Brown, H. M. (2001). Instrument. *Grove Music Online*.

Bryant, D. (2018). Cori spezzati in Composition and Sound. In K. Schiltz (Ed.), *A Companion to Music in Sixteenth-Century Venice*, pp. 371–394. Leiden, Netherlands: Brill.

Burgess, R. J. (2013). *The Art of Music Production: The Theory and Practice*. New York: Oxford University Press.

Burgess, R. J. (2014). *The History of Music Production*. New York: Oxford University Press.

Bush, D., and R. Kassel (Eds.) (2004). *The Organ: An Encyclopedia*. New York: Routledge.

Butler, M. J. (2014). *Playing with Something That Runs: Technology*. New York: Oxford Scholarship Online.

Buxton, W. (1986). Chunking and Phrasing and the Design of Human-Computer Dialogues. In H.-J. Kugler (Ed.), *Information Processing*, pp. 494–499. North-Holland Elsevier Science Publishers.

Bye, M. (2018). *MIDI Polyphonic Expression: A Bridge Between Acoustic and Electronic Instrument Experience?* Master's thesis, Universitetet i Oslo.

Cabe, P. A., and J. B. Pittenger (2000). Human sensitivity to acoustic information from vessel filling. *Journal of experimental psychology: Human Perception and Performance* 26(1), 313.

Cáceres, J.-P., and C. Chafe (2010). JackTrip: Under the hood of an engine for network audio. *Journal of New Music Research* 39(3), 183–187.

Cadoz, C. (1988). Instrumental gesture and musical composition. In *International Computer Music Conference*, pp. 1–12. Cologne, Germany: International Computer Music Association.

Cadoz, C., A. Luciani, J.-l. Florens, C. Roads, and F. Chadabe (1984). Responsive input devices and sound synthesis by stimulation of instrumental mechanisms: The cordis system. *Computer music journal* 8(3), 60–73.

Cadoz, C., and M. M. Wanderley (2000). Gesture-music. In M. M. Wanderley and M. Battier (Eds.), *Trends in Gestural Control of Music*, Volume 12, pp. 71–94. Paris: IRCAM.

Cage, J. (1961). *Silence: Lectures and Writings*. Middletown, CT: Wesleyan University Press.

Campbell, L., and M. Wanderley (2005). The Observation of Movement. MUMT 609 report, McGill University.

Camurri, A., B. Mazzarino, and G. Volpe (2005). A tool for analysis of expressive gestures: The EyesWeb Expressive Gesture Processing Library. *Ret 18*(2).

Cance, C. (2017). From Musical Instruments as Ontological Entities to Instrumental Quality: A Linguistic Exploration of Musical Instrumentality in the Digital Era. In *Musical Instruments in the 21st Century*, pp. 25–43. Singapore: Springer.

Carello, C., K. L. Anderson, and A. J. Kunkler-Peck (1998). Perception of object length by sound. *Psychological science 9*(3), 211–214.

Carelman, J. (1969). *Catalogue d'objets introuvables*. A. Paris: Balland.

Carr, I. (1992). *Keith Jarrett: The Man And His Music*. Cambridge, MA: Da Capo Press.

Chadabe, J. (1997). *Electric Sound: The Past and Promise of Electronic Music*. Upper Saddle River, NJ: Prentice Hall.

Chaigne, A., and J. Kergomard (2016). *Acoustics of Musical Instruments*. New York: Springer.

Chapman, J. (2015). *Emotionally Durable Design: Objects, Experiences and Empathy*. Oxon: Routledge.

Clark Estes, A. (2018). The First Great iPhone App Grows Up. *Gizmodo 3*(16).

Clarke, E. F. (2005). *Ways of Listening: An Ecological Approach to the Perception of Musical Meaning*. Oxford: Oxford University Press.

Clarke, E. F., and J. W. Davidson (1998). The Body in Performance. In *Composition-Performance Reception*, pp. 74–92. Aldershot, UK: Ashgate.

Clarke, E. F., and N. Cook (Eds.) (2004). *Empirical Musicology: Aims, Methods, Prospects*. New York: Oxford University Press.

Clayton, M., K. Jakubowski, T. Eerola, P. E. Keller, A. Camurri, G. Volpe, and P. Alborno (2020). Interpersonal Entrainment in Music Performance: Theory, Method, and Model. *Music Perception 38*(2), 136–194.

Clayton, M., R. Sager, and U. Will (2005). In time with the music: The concept of entrainment and its significance for ethnomusicology. In *European Meetings in Ethnomusicology*, Volume 1, pp. 3–75. Bucharest: Romanian Society for Ethnomusicology.

Clynes, M. (Ed.) (1982). *Music, Mind, and Brain: The Neuropsychology of Music*. New York: Springer Science & Business Media.

Collins, N. (2006). *Handmade Electronic Music: The Art of Hardware Hacking*. New York: Routledge.

Collins, N. (2009). *Handmade Electronic Music: The Art of Hardware Hacking* (Second ed.). New York: Routledge.

Collins, N. (2020). *Handmade Electronic Music: The Art of Hardware Hacking* (Third ed.). New York: Routledge.

Collins, N., A. McLean, J. Rohrhuber, and A. Ward (2003). Live coding in laptop performance. *Organised Sound 8*(3), 321–330.

Collins, N., and J. d'Escriván (2007). *The Cambridge Companion to Electronic Music*. Cambridge: Cambridge University Press.

Cook, P. R. (Ed.) (1999). *Music, Cognition, and Computerized Sound : An Introduction to Psychoacoustics*. Cambridge, MA: MIT Press.

Cook, P. R. (2001). Principles for Designing Computer Music Controllers. In I. Poupyrev, M. J. Lyons, S. S. Fels, and T. Blaine (Eds.), *Proceedings of the International Conference on New Interfaces for Musical Expression*, Seattle, pp. 3–6.

Cook, P. R. (2002). *Real Sound Synthesis for Interactive Applications*. Natick, MA: A K Peters.

Cooper, F., A. Liberman, and J. Borst (1951). The interconversion of audible and visible patterns as a basis for research in the perception of speech. *Proceedings of the National Academy of Sciences of the United States of America 37*(5), 318.

Cox, A. (2016). *Music and Embodied Cognition: Listening, Moving, Feeling, and Thinking*. Bloomington, IN: Indiana University Press.

Crawford, L., and W. D. Fastenow (2009). The Midi-AirGuitar , A serious Musical Controller with a Funny Name. In *Proceedings of the International Conference on New Interfaces for Musical Expression*, Pittsburgh, pp. 149–150.

Crowther, P. (2005). *Design For Disassembly Themes And Principles*. Melbourne: Royal Australian Institute of Architects, 1–7.

Dahl, S., and R. Bresin (2001). Is the player more influenced by the auditory than the tactile feedback from the instrument. In *Proc. of the COST-G6 Workshop on Digital Audio Effects (Dafx-01)*, Limerick, Ireland, pp. 194–197.

Dahlstedt, P. (2001). A MutaSynth in parameter space: Interactive composition through evolution. *Organised Sound 6*(02), 121–124.

Dahlstedt, P. (2017). Physical Interactions with Digital Strings - A hybrid approach to a digital keyboard instrument. In *Proceedings of the International Conference on New Interfaces for Musical Expression*, Copenhagen, pp. 115–120. Aalborg University Copenhagen.

Dainton, B. (2018). Temporal Consciousness. In E. N. Zalta (Ed.), *The Stanford Encyclopedia of Philosophy* (Winter 2018 ed.). Metaphysics Research Lab, Stanford University.

Danielsen, A., and I. Helseth (2016). Mediated Immediacy: The Relationship between Auditory and Visual Dimensions of Live Performance in Contemporary Technology-Based Popular Music. *Rock Music Studies 3*(1), 24–40.

Danielsen, A., K. Nymoen, E. Anderson, G. S. Câmara, M. T. Langerød, M. R. Thompson, and J. London (2019). Where is the beat in that note? Effects of attack, duration, and frequency on the perceived timing of musical and quasi-musical sounds. *Journal of Experimental Psychology: Human Perception and Performance 45*(3), 402–418.

Danielsen, A., M. R. Haugen, and A. R. Jensenius (2015). Moving to the Beat: Studying Entrainment to Micro-Rhythmic Changes in Pulse by Motion Capture. *Timing and Time Perception 3*, 133–154.

Dannenberg, R. B. (2018). Languages for Computer Music. *Frontiers in Digital Humanities 5*.

Dannenberg, R. B., and Z. Chi (2016). O2: Rethinking Open Sound Control. In *Proceedings of the International Computer Music Conference*, Utrecht, Netherlands, pp. 4.

Davidson, J. W., and J. S. Correia (2002). Body Movement. In R. Parncutt and G. E. McPherson (Eds.), *The Science and Psychology of Music Performance, Creative Strategies for Teaching and Learning*, pp. 237–250. New York: Oxford University Press.

de Haan, S., E. Rietveld, M. Stokhof, and D. Denys (2013). The phenomenology of deep brain stimulation-induced changes in OCD: An enactive affordance-based model. *Frontiers in Human Neuroscience 7*.

de Quay, Y., S. Skogstad, and A. R. Jensenius (2011). Dance Jockey: Performing Electronic Music by Dancing. *Leonardo Music Journal 21*, 11–12.

De Souza, J. (2017). *Music at Hand: Instruments, Bodies, and Cognition*. Oxford Studies in Music Theory. New York: Oxford University Press.

Dean, R. T., and A. McLean (Eds.) (2018). *The Oxford Handbook of Algorithmic Music*. Oxford: Oxford University Press.

Delalande, F. (1988). La gestique de Gould: éléments pour une sémiologie du geste musical. In G. Guertin (Ed.), *Glenn Gould Pluriel*, pp. 85–111. Verdun, Quebec: Louise Courteau.

DeNora, T. (1995). *Beethoven and the Construction of Genius*. Berkeley: University of California Press.

DeNora, T. (2000). *Music in Everyday Life*. Cambridge, UK: Cambridge University Press.

Devine, K. (2019). *Decomposed*. Cambridge, MA: MIT Press.

Dobrian, C., and D. Koppelman (2006). The E in NIME: Musical Expression with New Computer Interfaces. In N. Schnell, F. Bevilacqua, M. Lyons, and A. Tanaka (Eds.), *Proceedings of the International Conference on New Interfaces for Musical Expression*, Paris, pp. 277–282.

Dobrian, C., and F. Bevilacqua (2003). Gestural Control of Music Using the Vicon 8 Motion Capture System. In *Proceedings of the International Conference on New Interfaces for Musical Expression*, Montreal, pp. 161–163.

Donnarumma, M. (2016). *Configuring Corporeality: Performing Bodies, Vibrations and New Musical Instruments*. PhD diss., Goldsmiths, University of London.

Dourish, P. (2001). *Where the Action Is: The Foundations of Embodied Interaction*. Cambridge, MA: MIT Press.

Drioli, C., C. Allocchio, and N. Buso (2013). Networked performances and natural interaction via LOLA: Low latency high quality A/V streaming system. In *International Conference on Information Technologies for Performing Arts, Media Access, and Entertainment*, pp. 240–250. Springer.

Dudas, R. (2010). "Comprovisation": The Various Facets of Composed Improvisation within Interactive Performance Systems. *Leonardo Music Journal 20*, 29–31.

Eager, D., A.-M. Pendrill, and N. Reistad (2016). Beyond velocity and acceleration: Jerk, snap and higher derivatives. *European Journal of Physics 37*(6), 065008.

Edelman, G. M. (1987). *Neural Darwinism: The Theory of Neuronal Group Selection*. New York: Basic Books.

Eidsheim, N. S. (2017). Maria Callas's Waistline and the Organology of Voice. *The Opera Quarterly 33*(3-4), 249–268.

Eldridge, A., and C. Kiefer (2017). Self-resonating Feedback Cello: Interfacing gestural and generative processes in improvised performance. In *Proceedings of the International Conference on New Interfaces for Musical Expression*, Copenhagen, pp. 25–29. Aalborg University Copenhagen.

Ellerbrock, P. J., R. L. Grant, D. W. Konz, and J. P. Winkelmann (2004). Network device interface for digitally interfacing data channels to a controller via a network. US Patent US7552256B2.

Emerson, G., and H. Egermann (2018). Gesture-sound causality from the audience's perspective: Investigating the aesthetic experience of performances with digital musical instruments. *Psychology of Aesthetics, Creativity, and the Arts 12*(1), 96–109.

Enders, B. (2017). From Idiophone to Touchpad. The Technological Development to the Virtual Musical Instrument. In *Musical Instruments in the 21st Century*, pp. 45–58. Singapore: Springer.

Eno, B., and P. Chilvers (2008). Bloom. GenerativeMusic.com.

Erdem, Ç., Q. Lan, and A. R. Jensenius (2020). Exploring relationships between effort, motion, and sound in new musical instruments. *Human Technology: An Interdisciplinary Journal on Humans in ICT Environments 16*(3), 310–347.

Erdem, C., Q. Lan, J. Fuhrer, C. Martin, J. Tørresen, and A. R. Jensenius (2020). Towards Playing in the "Air": Modeling Motion–Sound Energy Relationships in Electric Guitar Performance Using Deep Neural Networks. In *Proceedings of the Sound and Music Computing Conference*. Birmingham, UK, pp. 177–183.

Essl, G. (2003). On gender in new music interface technology. *Organised Sound 8*(1), 19–30.

Evans, B. (2005). Foundations of a visual music. *Computer Music Journal 29*(4), 11–24.

Ewell, P. A. (2020). Music Theory and the White Racial Frame. *Music Theory Online 26*(2).

Fagereng, R. (2008). *Levende Elektronikk: En Studie Av Eivind Aarsets Bruk Av Gitar Og Elektronikk i Konsertsammenheng*. MA thesis, University of Oslo.

Favilla, S., J. Cannon, and G. Greenwood (2005). The Bent Leather Band Ensemble: Children of Grainger. In *Sound Scrips: Proceedings of the Inaugural Totally Huge New Musical Festival Conference*. Perth: Edith Cowan University.

Fels, S. S., and F. Vogt (2002). Tooka: Explorations of Two Person Instruments. In C. Casey, K. Schneider, and E. Hammond (Eds.), *Proceedings of the International Conference on New Interfaces for Musical Expression*, Dublin, pp. 50–55.

Fiebrink, R., D. Trueman, and P. R. Cook (2011). The Wekinator: Software for using machine learning to build real-time interactive systems. http://code.google.com/p/wekinator/. Accessed June 8, 2022.

Fiebrink, R., G. Wang, and P. R. Cook (2007). Don't Forget the Laptop: Using Native Input Capabilities for Expressive Musical Control. In *Proceedings of the International Conference on New Interfaces for Musical Expression*, New York, pp. 164–167.

Fiebrink, R., and L. Sonami (2020). Reflections on eight years of instrument creation with machine learning. In R. Michon and F. Schroeder (Eds.), *Proceedings of the International Conference on New Interfaces for Musical Expression*, Birmingham, UK, pp. 281–285. Birmingham City University.

Fiebrink, R., and P. R. Cook (2010). The Wekinator: A system for real-time, interactive machine learning in music. In *Proceedings of The International Society for Music Information Retrieval Conference*, Utrecht, Netherlands.

Fortuin, H. (1995). The Clavette: A Generalized Microtonal MIDI Keyboard Controller. In *Proceedings of the International Computer Music Conference.* Banff, Canada.

Freed, A. (2012). The Fingerphone: A Case Study of Sustainable Instrument Redesign. In G. Essl, B. Gillespie, M. Gurevich, and S. O'Modhrain (Eds.), *Proceedings of the International Conference on New Interfaces for Musical Expression*, Ann Arbor, MI. University of Michigan.

Freed, D. J. (1990). Auditory correlates of perceived mallet hardness for a set of recorded percussive sound events. *The Journal of the Acoustical Society of America 87*(1), 311–322.

Frid, E. (2019). *Diverse Sounds : Enabling Inclusive Sonic Interaction.* PhD diss., KTH Royal Institute of Technology, Stockholm.

Frisk, H. (2008). *Improvisation, Computers and Interaction Rethinking Human-Computer Interaction Through Music.* PhD diss., Malmö Academy of Music.

Frith, S., and S. Zagorski-Thomas (Eds.) (2012). *The Art of Record Production: An Introductory Reader for a New Academic Field.* Oxon: Routledge.

Fritz, C., J. Curtin, J. Poitevineau, and F.-C. Tao (2017). Listener evaluations of new and Old Italian violins. *Proceedings of the National Academy of Sciences 114*(21), 5395–5400.

Fritz, C., J. Curtin, J. Poitevineau, P. Morrel-Samuels, and F.-C. Tao (2012). Player preferences among new and old violins. *Proceedings of the National Academy of Sciences 109*(3), 760–763.

Gadir, T. (2017). Forty-Seven DJs, Four Women: Meritocracy, Talent, and Post-feminist Politics. *Dancecult 9*(1).

Gallese, V., L. Fadiga, L. Fogassi, and G. Rizzolatti (1996). Action recognition in the premotor cortex. *Brain 119*(2), 593–609.

Gang Qian, Feng Guo, T. Ingalls, L. Olson, J. James, and T. Rikakis (2004). A gesture-driven multimodal interactive dance system. In *2004 IEEE International*

Conference on Multimedia and Expo (ICME) (IEEE Cat. No.04TH8763), Taipei, Taiwan, pp. 1579–1582. IEEE.

Gaver, W. W. (1993a). How do we hear in the world? Explorations in ecological acoustics. *Ecological psychology 5*(4), 285–313.

Gaver, W. W. (1993b). What in the world do we hear? An ecological approach to auditory event perception. *Ecological psychology 5*(1), 1–29.

Gibet, S. (1987). *Codage, Representation et Traitement Du Geste Instrumental: Application a La Synthese de Sons Musicaux Par Simulation de Mecanismes Instrumentaux*. PhD diss., L'Institut National Polytechnique de Grenoble.

Gibson, J. J. (1977). The theory of affordances. In R. Shaw and J. Bransford (Eds.), *Perceiving, Acting, and Knowing: Toward an Ecological Psychology*, pp. 67–82. Hillsdale, NJ: Erlbaum.

Gibson, J. J. (1979). *The Ecological Approach to Visual Perception*. New York: Houghton-Mifflin.

Giordano, B. L., and S. McAdams (2006). Material identification of real impact sounds: Effects of size variation in steel, glass, wood, and plexiglass plates. *The Journal of the Acoustical Society of America 119*(2), 1171.

Godøy, R. I. (2001). Imagined Action, Excitation, and Resonance. In R. I. Godøy and H. Jørgensen (Eds.), *Musical Imagery*, pp. 237–250. Lisse, Netherlands: Swets and Zeitlinger.

Godøy, R. I. (2003). Motor-Mimetic Music Cognition. *Leonardo 36*(4), 317–319.

Godøy, R. I. (2006). Gestural-Sonorous Objects: Embodied extensions of Schaeffer's conceptual apparatus. *Organised Sound 11*(02), 149.

Godøy, R. I. (2008). Reflections on Chunking in Music. In A. Schneider (Ed.), *Systematic and Comparative Musicology: Concepts, Methods, Findings. Hamburger Jahrbuch Für Musikwissenschaft*, Volume 24, pp. 117–132. Vienna: Peter Lang.

Godøy, R. I. (2014). Understanding Coarticulation in Musical Experience. In M. Aramaki, O. Derrien, R. Kronland-Martinet, and S. Ystad (Eds.), *Sound, Music, and Motion*, Volume 8905, pp. 535–547. Cham, Switzerland: Springer.

Godøy, R. I., E. Haga, and A. R. Jensenius (2006a). Exploring Music-Related Gestures by Sound-Tracing - A Preliminary Study. In K. Ng (Ed.), *Proceedings of the COST287-ConGAS 2nd International Symposium on Gesture Interfaces for Multimedia Systems*, Leeds, UK, pp. 27–33.

Godøy, R. I., E. Haga, and A. R. Jensenius (2006b). Playing "air instruments": Mimicry of sound-producing gestures by novices and experts. In *Gesture in Human-Computer Interaction and Simulation*, pp. 256–267. Berlin: Springer.

Godøy, R. I., and H. Jorgensen (2001). *Musical Imagery*. New York: Taylor & Francis.

Goehr, L. (1992). *The Imaginary Museum of Musical Works : An Essay in the Philosophy of Music: An Essay in the Philosophy of Music*. Oxford: Clarendon Press.

González Sánchez, V., A. Zelechowska, and A. R. Jensenius (2018). Correspondences Between Music and Involuntary Human Micromotion During Standstill. *Frontiers in Psychology* 9(1382).

González Sánchez, V., A. Zelechowska, and A. R. Jensenius (2019). Analysis of the Movement-Inducing Effects of Music through the Fractality of Head Sway during Standstill. *Journal of Motor Behavior* 52(6), 734–749.

González Sánchez, V. E., C. P. Martin, A. Zelechowska, K. A. V. Bjerkestrand, V. Johnson, and A. R. Jensenius (2018). Bela-Based Augmented Acoustic Guitars for Sonic Microinteraction. In *Proceedings of the International Conference on New Interfaces for Musical Expression*, Blacksburg, VA, pp. 324–327. Virginia Tech.

Green, O. (2011). Agility and Playfulness: Technology and Skill in the Performance Ecosystem. *Organised Sound* 16(2), 134–44.

Grier, J. (2021). *Musical Notation in the West*. Cambridge: Cambridge University Press.

Grimshaw-Aagaard, M., M. Walther-Hansen, and M. Knakkergaard (2019). *The Oxford Handbook of Sound and Imagination*, Volume 2. New York: Oxford University Press.

Gritten, A., and E. King (Eds.) (2006). *Music and Gesture*. Hampshire, UK: Ashgate.

Gritten, A., and E. King (Eds.) (2011). *New Perspectives on Music and Gesture*. Hampshire, UK: Ashgate.

Guettler, K. (2002). *The Bowed String: On the Development of Helmholtz Motion and on the Creation of Anomalous Low Frequencies*. PhD diss., KTH Royal Institute of Technology, Stockholm.

Haga, E. (2008). *Correspondences between Music and Body Movement*. PhD diss., University of Oslo.

Haken, L., E. Tellman, and P. Wolfe (1998). An Indiscrete Music Keyboard. *Computer Music Journal* 22(1), 30.

Hansen, K. F. (2010). *The Acoustics and Performance of DJ Scratching – Analysis and Modeling*. PhD diss., KTH Royal Institute of Technology, Stockholm.

Hardcastle, W. J., and N. Hewlett (Eds.) (1999). *Coarticulation: Theory, Data, and Techniques*. Cambridge: Cambridge University Press.

Harper, A. (2011). *Infinite Music: Imagining the Next Millennium of Human Music-Making*. Winchester, UK: John Hunt Publishing.

Hasse, S. (2012). I am Sitting in a Room. *Body, Space & Technology 11*(0).

Hatten, R. S. (2004). *Interpreting Musical Gestures, Topics, and Tropes: Mozart, Beethoven, Schubert*. Bloomington, IN: Indiana University Press.

Haueisen, J., and T. R. Knösche (2001). Involuntary motor activity in pianists evoked by music perception. *Journal of cognitive neuroscience* 13(6), 786–792.

Haugen, M. R. (2016). *Music-Dance. Investigating Rhythm Structures in Brazilian Samba and Norwegian Telespringar Performance.* Phd diss., University of Oslo.

Haynes, B. (2002). *A History of Performing Pitch: The Story of "A"*. Lanham, MD: Scarecrow Press.

Hazas, M., and L. Nathan (2017). *Digital Technology and Sustainability: Engaging the Paradox.* Oxon: Routledge.

Heller, F., I. M. C. Ruiz, and J. Borchers (2017). An Augmented Flute for Beginners. In *Proceedings of the International Conference on New Interfaces for Musical Expression*, Copenhagen, pp. 34–37. Aalborg University Copenhagen.

Heyde, H. (1975). *Grundlagen des natürlichen Systems der Musikinstrumente.* Leipzig, Germany: VEB Deutscher Verlag für Musik.

Hickok, G., B. Buchsbaum, C. Humphries, and T. Muftuler (2003). Auditory–motor interaction revealed by fMRI: Speech, music, and working memory in area Spt. *Journal of Cognitive Neuroscience 15*(5), 673–682.

Holland, S., T. Mudd, K. Wilkie-McKenna, A. McPherson, and M. M. Wanderley (2019). *New Directions in Music and Human-Computer Interaction.* Cham, Switzerland: Springer.

Holmes, T. (2002). *Electronic and Experimental Music: Pioneers in Technology and Composition* (Second ed.). New York: Routledge.

Holmes, T. (2012). *Electronic and Experimental Music: Technology, Music, and Culture* (Fourth ed.). New York: Routledge.

Holmes, T. (2016). *Electronic and Experimental Music: Technology, Music, and Culture* (Fifth ed.). New York: Routledge.

Holopainen, R. (2012). *Self-Organised Sound with Autonomous Instruments: Aesthetics and Experiments.* PhD diss., University of Oslo.

Howard, D. M., and J. A. S. Angus (2007). *Acoustics and Psychoacoustics* (Third ed.). Oxford: Focal Press.

Hunt, A., and M. M. Wanderley (2002). Mapping performer parameters to synthesis engines. *Organised Sound* 7(2), 97–108.

Hunt, A., M. M. Wanderley, and M. Paradis (2003). The importance of parameter mapping in electronic instrument design. *Journal of New Music Research 32*(4), 429–440.

Hurley, S. (1998). *Consciousness in Action.* Cambridge, MA: Harvard University Press.

Husserl, E. (1991). *On the Phenomenology of the Consciousness of Internal Time (1893–1917).* Dordrecht, Netherlands: Kluwer Academic Publishers.

Huston, R. (2008). *Principles of Biomechanics.* Boca Raton, FL: CRC Press.

Iñesta, J. M., D. C. Conklin, R. Ramírez-Melendez, and T. M. Fiore (2018). *Machine Learning and Music Generation.* Oxon: Routledge.

Jack, R., J. Harrison, and A. McPherson (2020). Digital musical instruments as research products. In *Proceedings of the International Conference on New Interfaces for Musical Expression*, Birmingham, UK, pp. 530–536.

Jacob, S. (2019). *Acousmatic Experience and Musical Movement: A Pluralistic Conception*. PhD diss., Durham University.

Jarrett, K. (1991). *The Köln Concert*. Mainz, Germany: Schott.

Jawad, K. (2020). *Gatekeepers by Design? Gender HCI for Audio and Music Hardware*. PhD diss., Norwegian University of Science and Technology.

Jawad, K., and A. Xambó (2020). How to Talk of Music Technology: An Interview Analysis Study of Live Interfaces for Music Performance among Expert Women. In *Proceedings of the International Conference on Live Interfaces*, Trondheim, Norway, pp. 41–47.

Jenkins, M. (2007). *Analog Synthesizers - Understanding, Performing, Buying: From the Legacy of Moog to Software Synthesis*. Burlington, MA: Taylor & Francis.

Jensenius, A. R. (2007). *Action–Sound: Developing Methods and Tools to Study Music-Related Body Movement*. PhD diss., University of Oslo.

Jensenius, A. R. (2013a). An Action-Sound Approach to Teaching Interactive Music. *Organised Sound 18*(2), 178–189.

Jensenius, A. R. (2013b). Kinectofon: Performing with Shapes in Planes. In *Proceedings of the International Conference on New Interfaces for Musical Expression*, Daejeon, Korea.

Jensenius, A. R. (2013c). Some video abstraction techniques for displaying body movement in analysis and performance. *Leonardo 46*(1), 53–60.

Jensenius, A. R. (2014). To Gesture or Not? An Analysis of Terminology in NIME Proceedings 2001-2013. In *Proceedings of the International Conference on New Interfaces for Musical Expression*, London, pp. 217–220.

Jensenius, A. R. (2017a). Exploring music-related micromotion. In C. Wöllner (Ed.), *Body, Sound and Space in Music and Beyond: Multimodal Explorations*, pp. 29–48. Oxon: Routledge.

Jensenius, A. R. (2017b). Sonic Microinteraction in "the Air." In M. Lesaffre, P.-J. Maes, and M. Leman (Eds.), *The Routledge Companion to Embodied Music Interaction*, pp. 431–439. New York: Routledge.

Jensenius, A. R. (2018). Methods for studying music-related body motion. In R. Bader (Ed.), *Handbook of Systematic Musicology*, pp. 567–580. Berlin: Springer-Verlag.

Jensenius, A. R., and A. Voldsund (2012). The Music Ball Project: Concept, Design, Development, Performance. In *Proceedings of the International Conference on New Interfaces For Musical Expression*, Ann Arbor, MI, pp. 300–303.

Jensenius, A. R., A. Zelechowska, and V. E. Gonzalez-Sanchez (2017). The Musical Influence on People's Micromotion when Standing Still in Groups. In *Proceedings of the Sound and Music Computing Conference*, Helsinki, pp. 195–200.

Jensenius, A. R., and K. A. V. Bjerkestrand (2012). Exploring micromovements with motion capture and sonification. In A. L. Brooks (Ed.), *Arts and Technology, Revised Selected Papers*, Volume 101 of *LNICST*, pp. 100–107. Berlin: Springer.

Jensenius, A. R., K. Nymoen, S. Skogstad, and A. Voldsund (2012). A Study of the Noise-Level in Two Infrared Marker-Based Motion Capture Systems. In *Proceedings of the Sound and Music Computing Conference*, Copenhagen, pp. 258–263.

Jensenius, A. R., and M. J. Lyons (Eds.) (2017). *A NIME Reader: Fifteen Years of New Interfaces*. Berlin: Springer.

Jensenius, A. R., M. M. Wanderley, R. I. Godøy, and M. Leman (2010). Musical gestures: Concepts and methods in research. In *Musical Gestures: Sound, Movement, and Meaning*, pp. 12–35. New York: Routledge.

Jensenius, A. R., and R. I. Godøy (2013). Sonifying the shape of human body motion using motiongrams. *Empirical Musicology Review 8*(2), 73–83.

Jensenius, A. R., R. Koehly, and M. M. Wanderley (2006). Building Low-Cost Music Controllers. In *CMMR 2005*, Volume 3902 of *LNCS*, pp. 123–129. Berlin: Springer.

Jensenius, A. R., T. Kvifte, and R. I. Godøy (2006). Towards a Gesture Description Interchange Format. In *Proceedings of the International Conference on New Interfaces for Musical Expression*, Paris, pp. 176–179.

Jensenius, A. R., V. E. González Sánchez, A. Zelechowska, and K. A. V. Bjerkestrand (2017). Exploring the Myo controller for sonic microinteraction. In *Proceedings of the International Conference on New Interfaces for Musical Expression*, Copenhagen, pp. 442–445.

Johnson, P. (1997). Musical Works, Musical Performances. *The Musical Times 138*(1854), 4.

Johnstone, E. (1985). The Rolky: A Poly-Touch Controller for Electronic Music. In *Proceedings of the International Computer Music Conference*, Burnaby, Canada, pp. 291–294.

Jones, E. (2020). *DIY Music and the Politics of Social Media*. New York: Bloomsbury.

Jones, R., P. Driessen, A. Schloss, and G. Tzanetakis (2009). A Force-Sensitive Surface for Intimate Control. In *Proceedings of the International Conference on New Interfaces for Musical Expression*, Pittsburgh, pp. 236–241.

Jordà, S. (2005). *Digital Lutherie : Crafting Musical Computers for New Musics' Performance and Improvisation*. PhD diss., Universitat Pompeu Fabra.

Jordà, S., M. Kaltenbrunner, G. Geiger, and R. Bencina (2005). The reacTable*. In *Proceedings of the 2005 International Computer Music Conference*, Barcelona, pp. 579–582. San Francisco: ICMA.

Jordan, W. E. (1953). Norman McLaren: His Career and Techniques. *The Quarterly of Film Radio and Television 8*(1), 1–14.

Jostmann, N. B., D. Lakens, and T. W. Schubert (2009). Weight as an Embodiment of Importance. *Psychological Science 20*(9), 1169–1174.

Jota, R., A. Ng, P. Dietz, and D. Wigdor (2013). How Fast is Fast Enough? A Study of the Effects of Latency in Direct-Touch Pointing Tasks. In *Proceedings of the ACM Conference on Human Factors in Computing Systems (CHI)*, Paris, pp. 2291–2300.

Jungleib, S. (1996). *General MIDI*. Middleton, WI: A-R Editions.

Kaltenbrunner, M., T. Bovermann, R. Bencina, and E. Costanza (2005). TUIO - A Protocol for Table-Based Tangible User Interfaces. In *Proceedings of the 6th International Gesture Workshop*, Vannes, France.

Karjalainen, M., T. Mäki-Patola, A. Kanerva, and A. Huovilainen (2006). Virtual air guitar. *Journal of the Audio Engineering Society 54*(10), 964–980.

Karplus, K., and A. Strong (1983). Digital synthesis of plucked-string and drum timbres. *Computer Music Journal* 7(2), 43–55.

Katz, M. (2010). *Capturing Sound: How Technology Has Changed Music*. Berkeley: University of California Press.

Keele, S. W. (1973). *Attention and Human Performance*. Pacific Palisades, CA: Goodyear Publishing Company.

Keislar, D. (1987). History and Principles of Microtonal Keyboards. *Computer Music Journal 11*(1), 18–28.

Keislar, D. (2009). A Historical View of Computer Music Technology. In R. T. Dean (Ed.), *The Oxford Handbook of Computer Music*. New York: Oxford University Press.

Kelkar, T. (2019). *Computational Analysis of Melodic Contour and Body Movement*. PhD diss., University of Oslo.

Kelkar, T., and A. R. Jensenius (2018). Analyzing Free-Hand Sound-Tracings of Melodic Phrases. *Applied Sciences 8*(1), 135.

Kendon, A. (1980). Gesticulation and speech: Two aspects of the process of utterance. In M. R. Key (Ed.), *The Relationship between Verbal and Nonverbal Communication*, pp. 207–227. The Hague, Netherlands: Mouton Publishers.

Kendon, A. (2004). *Gesture: Visible Action as Utterance*. Cambridge: Cambridge University Press.

Kerman, J. (1985). *Contemplating Music: Challenges to Musicology*. Cambridge, MA: Harvard University Press.

Kimura, M., N. Rasamimanana, F. Bevilacqua, N. Schnell, B. Zamborlin, and E. Fléty (2012). Extracting Human Expression For Interactive Composition with the Augmented Violin. In *Proceedings of the International Conference on New Interfaces for Musical Expression*, Ann Arbor, MI. University of Michigan.

King, S., and K. Chang (2016). *Understanding Industrial Design: Principles for UX and Interaction Design*. Sebastopol, CA: O'Reilly Media, Inc.

Kjus, Y., and A. Danielsen (2016). Live mediation: Performing concerts using studio technology. *Popular Music 35*(3), 320–337.

Knapp, R. B., and H. S. Lusted (1990). A Bioelectric Controller for Computer Music Applications. *Computer Music Journal 14*(1), 42–47.

Knotts, S., and N. Collins (2014). The Politics of Laptop Ensembles: A Survey of 160 Laptop Ensembles and their Organisational Structures. In B. Caramiaux, K. Tahiroglu, R. Fiebrink, and A. Tanaka (Eds.), *Proceedings of the International Conference on New Interfaces for Musical Expression*, London, pp. 191–194. Goldsmiths, University of London.

Kockelman, P. (2013). *Agent, Person, Subject, Self: A Theory of Ontology, Interaction, and Infrastructure*. New York: Oxford University Press.

Koehly, R. (2011). *Fabrication of Sustainable Resistive-Based Paper Touch Sensors: Application to Music Technology*. PhD diss., McGill University.

Kohler, E., C. Keysers, M. A. Umilta, L. Fogassi, V. Gallese, and G. Rizzolatti (2002). Hearing Sounds, Understanding Actions: Action Representation in Mirror Neurons. *Science* 297(5582), 846–848.

Kozak, M. (2019). *Enacting Musical Time: The Bodily Experience of New Music*. New York: Oxford University Press.

Kramer, J. D. (1988). *The Time of Music: New Meanings, New Temporalities, New Listening Strategies*. New York: Schirmer Books.

Kreiman, J., and D. Sidtis (2011). *Foundations of Voice Studies: An Interdisciplinary Approach to Voice Production and Perception*. Malden, MA: John Wiley & Sons.

Kunkler-Peck, A. J., and M. T. Turvey (2000). Hearing shape. *Journal of Experimental psychology: Human Perception and Performance* 26(1), 279.

Kurtenbach, G., and E. A. Hulteen (1990). Gestures in human-computer communication. In B. Laurel and S. J. Mountford (Eds.), *The Art of Human-Computer Interface Design*. Reading, MA: Addison-Wesley.

Kvifte, T. (1989). *Instruments and the Electronic Age: Toward a Terminology for a Unified Description of Playing Technique*. Oslo: Solum Forlag.

Kvifte, T. (2008). What is a musical instrument. *Svensk tidskrift för musikforskning 1*(2008), 45–56.

Kvifte, T., and A. R. Jensenius (2006). Towards a Coherent Terminology and Model of Instrument Description and Design. In *Proceedings of the International Conference on New Interfaces for Musical Expression*, pp. 220–225.

Lahav, A., E. Saltzman, and G. Schlaug (2007). Action Representation of Sound: Audiomotor Recognition Network While Listening to Newly Acquired Actions. *Journal of Neuroscience* 27(2), 308–314.

Lähdeoja, O. (2015). An Augmented Guitar With Active Acoustics. In *Proceedings of the Sound and Music Computing Conference*, Maynooth, Ireland.

Lamb, R., and A. Robertson (2011). Seaboard: A New Piano Keyboard-related Interface Combining Discrete and Continuous Control. In *Proceedings of the International Conference on New Interfaces for Musical Expression*, Oslo, pp. 503–506.

Langer, S. K. (1957). *Philosophy in a New Key: A Study in the Symbolism of Reason, Rite, and Art*. Cambridge, MA: Harvard University Press.

Lartillot, O., K. Nymoen, G. S. Câmara, and A. Danielsen (2021). Computational localization of attack regions through a direct observation of the audio waveform. *The Journal of the Acoustical Society of America 149*(1), 723–736.

Launay, J. (2015). Musical Sounds, Motor Resonance, and Detectable Agency. *Empirical Musicology Review 10*(1-2), 30–40.

Launay, J., R. T. Dean, and F. Bailes (2016). Rapid learning of associations between sound and action through observed movement. A TMS study. *Psychomusicology 26*(1), 35–42.

Leante, L. (2014). Gesture and imagery in music performance: Perspectives from North Indian classical music. In T. Shephard and A. Leonard (Eds.), *The Routledge Companion to Music and Visual Culture*, pp. 145–152. New York: Routledge.

Lee, D. (2019). Hornbostel-Sachs Classification. *Encyclopedia of Knowledge Organization*. Edmonton, Canada: ISKO.

Leeuw, H. (2012). The electrumpet, additions and revisions. In *Proceedings of the International Conference on New Interfaces for Musical Expression*, Ann Arbor. University of Michigan.

Leman, M. (2008). *Embodied Music Cognition and Mediation Technology*. Cambridge, MA: MIT Press.

Leman, M. (2016). *The Expressive Moment: How Interaction (with Music) Shapes Human Empowerment*. Cambridge, MA: MIT Press.

Lepri, G., and A. McPherson (2019). Fictional Instruments, Real Values: Discovering Musical Backgrounds with Non-Functional Prototypes. In *Proceedings of the International Conference on New Interfaces for Musical Expression*, Porto Alegre, Brazil.

Lesaffre, M., P.-J. Maes, and M. Leman (Eds.) (2017). *The Routledge Companion to Embodied Music Interaction*. New York: Routledge.

Levitin, D. J. (2014). Expert violinists can't tell old from new. *Proceedings of the National Academy of Sciences 111*(20), 7168–7169.

Levitin, D. J., S. McAdams, and R. L. Adams (2002). Control parameters for musical instruments: A foundation for new mappings of gesture to sound. *Organised Sound 7*(2), 171–189.

Liberman, A. M., and I. G. Mattingly (1985). The motor theory of speech perception revised. *Cognition 21*(1), 1–36.

Libin, L. (2018). Musical instrument. *Grove Music Online*. Oxford: Oxford University Press. doi:10.1093/omo/9781561592630.013.3000000097.

Lindblom, B. (1991). The Status of Phonetic Gestures. In I. M. Mattingly and M. Studdert-Kennedy (Eds.), *Modularity and the Motor Theory of Speech Perception, Proceedings of a Conference to Honor Alvin M. Liberman*, pp. 7–24. Florence, KY: Lawrence Erlbaum Associates.

Linn, R. (2013). LinnStrument and other new expressive musical controllers. *The Journal of the Acoustical Society of America 134*(5), 4053–4053.

London, J. (2004). *Hearing in Time: Psychological Aspects of Musical Meter*. New York: Oxford University Press.

Lorenz, E. N. (1963). Deterministic nonperiodic flow. *Journal of the atmospheric sciences 20*(2), 130–141.

Lossius, T. (2007). *Sound – Space – Body : Reflections on Artistic Practice*. Research Fellowship in the Arts thesis, Bergen National Academy of the Arts, Bergen, Norway.

Loughridge, D., and T. Patteson (2013). Museum of Imaginary Musical Instruments. http://imaginaryinstruments.org/about/. Accessed March 1, 2022.

Loy, G. (1985). Musicians make a standard: The MIDI phenomenon. *Computer Music Journal 9*(4), 8–26.

Lutfi, R. A. (2001). Auditory detection of hollowness. *The Journal of the Acoustical Society of America 110*(2), 1010.

Machover, T. (2004). Shaping minds musically. *BT Technology Journal* 22(4), 171–179.

Maeda, J. (2006). *The Laws of Simplicity: Design, Technology, Business, Life*. Cambridge, MA: MIT Press.

Magnusson, T. (2009). Of epistemic tools: Musical instruments as cognitive extensions. *Organised Sound 14*(2), 168–176.

Magnusson, T. (2010). Designing Constraints: Composing and Performing with Digital Musical Systems. *Computer Music Journal 34*(4), 62–73.

Magnusson, T. (2017). Musical Organics: A Heterarchical Approach to Digital Organology. *Journal of New Music Research 46*(3), 286–303.

Magnusson, T. (2018). Ergomimesis: Towards a language describing instrumental transductions. In *Proceedings: ICLI 2018, 4th International Conference on Live Interfaces*, pp. 78–85. Porto, Portugal: Universidade do Porto.

Magnusson, T. (2019). *Sonic Writing: Technologies of Material, Symbolic and Signal Inscriptions*. New York: Bloomsbury Academic.

Magnusson, T. (2021). The migration of musical instruments: On the socio-technological conditions of musical evolution. *Journal of New Music Research 50*(2), 175–183.

Mahillon, V.-C. (1880). *Catalogue Descriptif & Analytique Du Musée Instrumental Du Conservatoire Royal de Bruxelles: Précédé d'un Essai de Classification de Tous Les Instruments Anciens et Modernes*. Gand (Ghent) Belgium: C. Annoot-Braeckman.

Mankoff, J. C., E. Blevis, A. Borning, A. Borning, B. Friedman, S. R. Fussell, J. Hasbrouck, A. Woodruff, and P. Sengers (2007). Environmental sustainability and interaction. In *CHI'07 Extended Abstracts on Human Factors in Computing Systems*, pp. 2121–2124. San Jose, CA: ACM.

Mann, S. (2007). Natural Interfaces for Musical Expression : Physiphones and a Physics-Based Organology. In *Proceedings of the International Conference on New Interfaces for Musical Expression*, New York City, pp. 118–123.

Marino, G., M.-H. Serra, J.-M. Raczinski, G. Marino, and M.-H. Serra (1993). The UPIC System: Origins and Innovations. *Perspectives of New Music 31*(1), 258.

Marshall, M. T., and M. M. Wanderley (2006). Evaluation of Sensors as Input Devices for Computer Music Interfaces. In R. Kronland-Martinet, T. Voinier, and S. Ystad (Eds.), *Computer Music Modeling and Retrieval*, Volume 3902, pp. 130–139. Berlin: Springer.

Martin, C. P., A. R. Jensenius, and J. Torresen (2018). Composing an Ensemble Standstill Work for Myo and Bela. In *Proceedings of the International Conference on New Interfaces for Musical Expression*, Blacksburg, VA, pp. 196–197. Virginia Tech.

Masu, R., A. P. Melbye, J. Sullivan, and A. R. Jensenius (2021). NIME and the Environment: Toward a More Sustainable NIME Practice. In *Proceedings of the International Conference on New Interfaces for Musical Expression.* Shanghai, China.

Mayton, B., G. Dublon, N. Joliat, and J. A. Paradiso (2012). Patchwork: Multi-User Network Control of a Massive Modular Synthesizer. In *Proceedings of the International Conference on New Interfaces for Musical Expression*, Ann Arbor. University of Michigan.

McAdams, S. (2000). The psychomechanics of real and simulated sound sources. *The Journal of the Acoustical Society of America 107*, 2792.

McCaleb, D. J. M. (2014). *Embodied Knowledge in Ensemble Performance.* Farnham, UK: Ashgate.

McClary, S. (1991). *Feminine Endings: Music, Gender, and Sexuality.* Minneapolis, MN: University of Minnesota Press.

McDonnell, M. (2020). *Finding Visual Music in Its 20th-Century History.* PhD diss., University of Dublin.

McGurk, H., and J. MacDonald (1976). Hearing lips and seeing voices. *Nature 264*(23/30), 746–748.

McNeill, D. (1992). *Hand and Mind: What Gestures Reveal About Thought.* Chicago: University of Chicago Press.

McNeill, D. (2005). *Gesture and Thought.* Chicago: University of Chicago Press.

McPherson, A. (2012). TouchKeys: Capacitive Multi-Touch Sensing on a Physical Keyboard. In G. Essl, B. Gillespie, M. Gurevich, and S. O'Modhrain (Eds.), *Proceedings of the International Conference on New Interfaces for Musical Expression*, Ann Arbor. University of Michigan.

McPherson, A., and K. Tahiroglu (2020). Idiomatic Patterns and Aesthetic Influence in Computer Music Languages. *Organised Sound: an international journal of music and technology 25*(1), 53–63.

McPherson, A., R. Jack, and G. Moro (2016). Action-Sound Latency: Are Our Tools Fast Enough? In *Proceedings of the International Conference on New Interfaces for Musical Expression*, Volume 16 of *2220-4806*, Brisbane, Australia, pp. 20–25. Queensland Conservatorium Griffith University.

McPherson, A., and V. Zappi (2015). An environment for submillisecond-latency audio and sensor processing on BeagleBone Black. In *Proceedings of the AES 138th Convention*, Warsaw.

Medeiros, C. B., and M. M. Wanderley (2014). A Comprehensive Review of Sensors and Instrumentation Methods in Devices for Musical Expression. *Sensors 14*(8), 13556–13591.

Mehrabian, A., and S. L. Friedman (1986). An analysis of fidgeting and associated individual differences. *Journal of Personality 54*(2), 406–429.

Melbye, A. P., and H. A. Ulfarsson (2020). Sculpting the behaviour of the Feedback-Actuated Augmented Bass: Design strategies for subtle manipulations of string feedback using simple adaptive algorithms. In *Proceedings of the International Conference on New Interfaces for Musical Expression*, Birmingham, UK, pp. 259–266.

Menin, D., and A. Schiavio (2012). Rethinking musical affordances. *Avant 3*(2), 201–215.

Métois, E. (1997). *Musical Sound Information: Musical Gestures and Embedding Synthesis*. PhD diss., Massachusetts Institute of Technology.

Miller, G. A. (1956). The magical number seven, plus or minus two: Some limits on our capacity for processing information. *Psychological Review 63*, 81–97.

MIMO, C. (2011). Revision of the Hornbostel-Sachs Classification of Musical Instruments by the MIMO Consortium. Technical report, MIMO Consortium.

Miranda, E. R. (2021). *Handbook of Artificial Intelligence for Music: Foundations, Advanced Approaches, and Developments for Creativity*. Cham, Switzerland: Springer.

Miranda, E. R., and M. M. Wanderley (2006). *New Digital Musical Instruments: Control and Interaction beyond the Keyboard*. Middleton, WI: A-R Editions.

Mithen, S. J. (2006). *The Singing Neanderthals: The Origins of Music, Language, Mind, and Body*. Cambridge, MA: Harvard University Press.

Moeslund, T. B., A. Hilton, and V. Krüger (2006). A survey of advances in vision-based human motion capture and analysis. *Computer Vision and Image Understanding 104*(2-3), 90–126.

Moeslund, T. B., and E. Granum (2001). A Survey of Computer Vision-Based Human Motion Capture. *Computer Vision and Image Understanding 81*(3), 231–268.

Momeni, A., and D. Wessel (2003). Characterizing and Controlling Musical Material Intuitively with Geometric Models. In M. M. Wanderley, R. McKenzie, and L. Ostiguy (Eds.), *Proceedings of the International Conference on New Interfaces for Musical Expression*, Montreal, pp. 54–62.

Moog, R. A. (1982). A Multiply Touch-Sensitive Clavier for Computer Music Systems. In *International Computer Music Conference*, Venice, Italy, pp. 601–605.

Moog, R. (1988). The Musician: Alive and well in the world of electronics. In *The Biology of music making: Proceedings of the 1984 Denver conference*. St. Louis: MMB Music, pp. 214–220.

Moore, A. (2016). *Sonic Art: An Introduction to Electroacoustic Music Composition*. New York: Routledge.

Moore, F. R. (1988). The dysfunctions of MIDI. *Computer music journal 12*(1), 19–28.

Morreale, F. (2021). Where Does the Buck Stop? Ethical and Political Issues with AI in Music Creation. *Transactions of the International Society for Music Information Retrieval 4*(1), 105–113.

Mulder, A., S. Fels, and K. Mase (1997). Mapping virtual object manipulation to sound variation. In T. Rai and R. Basset (Eds.), *IPSJ SIG Notes*, Tokyo, pp. 63–68.

Murray-Rust, D., and A. Smaill (2011). Towards a model of musical interaction and communication. *Artificial Intelligence 175*(9), 1697–1721.

Neuhoff, J. G. (2001). Perceiving acoustic source orientation in three-dimensional space. In *Proceedings of the International Conference on Auditory Display*, Espoo, Finland, pp. 231–234.

Noë, A. (2004). *Action in Perception*. Cambridge, MA: MIT Press.

Norman, A. D. (1988). *The Design of Everyday Things*. New York: Basic Books.

Norman, D. (2013). *The Design of Everyday Things: Revised and Expanded Edition*. New York: Basic Books.

Norman, D. A. (2004). *Emotional Design: Why We Love (or Hate) Everyday Things*. New York: Basic Books.

Nymoen, K. (2013). *Methods and Technologies for Analysing Links Between Musical Sound and Body Motion*. PhD diss., University of Oslo.

Nymoen, K., A. Chandra, K. Glette, and J. Torresen (2014). Decentralized harmonic synchronization in mobile music systems. In *International Conference on Awareness Science and Technology (iCAST)*, Paris, pp. 1–6. IEEE.

Nymoen, K., A. Voldsund, S. Skogstad, A. R. Jensenius, and J. Torresen (2012). Comparing Motion Data from an iPod Touch to a High-End Optical Infrared Marker-Based Motion Capture System. In *Proceedings of the International Conference on New Interfaces for Musical Expression*, Ann Arbor, MI. University of Michigan.

Nymoen, K., M. R. Haugen, and A. R. Jensenius (2015). MuMYO - Evaluating and Exploring the MYO Armband for Musical Interaction. In *Proceedings of the International Conference on New Interfaces For Musical Expression*, Baton Rouge, LA, pp. 215–218.

Nymoen, K., S. Skogstad, and A. R. Jensenius (2011). SoundSaber: A Motion Capture Instrument. In *Proceedings of the International Conference on New Interfaces for Musical Expression*, Oslo, pp. 312–315.

Oh, J., J. Herrera, N. J. Bryan, L. Dahl, and G. Wang (2010). Evolving the mobile phone orchestra. In *Proceedings of the International Conference on New Interfaces for Musical Expression*, Sydney, Australia, pp. 82–87.

Ojanen, M., J. Suominen, T. Kallio, and K. Lassfolk (2007). Design Principles and User Interfaces of Erkki Kurenniemi's Electronic Musical Instruments of the 1960's and 1970's. In *Proceedings of the International Conference on New Interfaces for Musical Expression*, New York, pp. 88–93.

Oliveros, P., S. Weaver, M. Dresser, J. Pitcher, J. Braasch, and C. Chafe (2009). Telematic music: Six perspectives. *Leonardo Music Journal 19*(1), 95–96.

Olsen, D. A. (1986). Is it time for another-phone? *SEM Newsletter 20*(4), 5.

O'Neill, S. (2008). *Interactive Media: The Semiotics of Embodied Interaction.* London: Springer.

Paine, G. (2015). Interaction as material: The techno-somatic dimension. *Organised Sound 20*(01), 82–89.

Palacio-Quintin, C. (2008). Eight Years of Practice on the Hyper-Flute: Technological and Musical Perspectives. In *Proceedings of the International Conference on New Interfaces for Musical Expression*, Genoa, Italy, pp. 293–298.

Papanek, V. (1985). *Design for the Real World: Human Ecology and Social Change.* Chicago: Academy Chicago Publishers.

Papetti, S., and C. Saitis (Eds.) (2018). *Musical Haptics.* Springer Series on Touch and Haptic Systems. Cham, Switzerland: Springer.

Paradiso, J., C. Abler, K.-y. Hsiao, and M. Reynolds (1997). The magic carpet: Physical sensing for immersive environments. In *CHI'97 Extended Abstracts on Human Factors in Computing Systems*, pp. 277–278. Atlanta, GA: ACM.

Parncutt, R. (2007). Systematic Musicology and the History and Future of Western Musical Scholarship. *Journal of Interdisciplinary Music Studies 1*(1), 1–32.

Peress, M. (2004). *Dvorak to Duke Ellington: A Conductor Explores America's Music and Its African American Roots.* New York: Oxford University Press.

Pérez, M. A. O., and R. B. Knapp (2008). BioTools: A Biosignal Toolbox for Composers and Performers. In R. Kronland-Martinet, S. Ystad, and K. Jensen (Eds.), *Computer Music Modeling and Retrieval. Sense of Sounds*, Number 4969 in Lecture Notes in Computer Science, pp. 441–452. Berlin: Springer.

Picard, R. W. (1997). *Affective Computing.* Cambridge, MA: MIT Press.

Rautaray, S. S., and A. Agrawal (2015). Vision-based hand gesture recognition for human computer interaction: A survey. *Artificial Intelligence Review 43*(1), 1–54.

Reboursière, L., C. Frisson, O. Lähdeoja, J. A. Mills, C. Picard-Limpens, and T. Todoroff (2010). Multimodal Guitar : A Toolbox For Augmented Guitar Performances. In *Proceedings of the International Conference on New Interfaces for Musical Expression*, Sydney, pp. 415–418.

Repp, B. H. (1987). The sound of two hands clapping: An exploratory study. *The Journal of the Acoustical Society of America 81*(4), 1100–1109.

Rietveld, E., and J. Kiverstein (2014). A Rich Landscape of Affordances. *Ecological Psychology 26*(4), 325–352.

Risset, J.-C. (2015). Recollections and Reflections on Organised Sound. *Organised Sound 20*(1), 15–22.

Rizzolatti, G., and M. A. Arbib (1998). Language within our grasp. *Trends in Neuroscience 21*, 188–194.

Roads, C. (1996). *The Computer Music Tutorial*. Cambridge, MA: MIT Press.

Rocchesso, D., and F. Fontana (2003). *The Sounding Object*. Firenze, Italy: Mondo estremo.

Roehmann, F. L., F. R. Wilson, and Biology of Music Making Conference (Eds.) (1988). *The Biology of Music Making: Proceedings of the 1984 Denver Conference*. St. Louis: MMB Music.

Rosch, E., C. B. Mervis, W. D. Gray, D. M. Johnson, and P. Boyes-Braem (1976). Basic Objects in Natural Categories. *Cognitive Psychology 8*(3), 382–439.

Rossing, T. D., F. R. Moore, and P. A. Wheeler (2002). *The Science of Sound* (Third ed.). Boston, MA: Addison Wesley.

Rothstein, J. (1992). *MIDI: A Comprehensive Introduction*. Number 7 in The Computer Music and Digital Audio Series. Madison, WI: A-R Ed.

Rowe, R. (1993). *Interactive Music Systems: Machine Listening and Composing*. Cambridge, MA: MIT Press.

Rowe, R. (2001). *Machine Musicianship*. Cambridge, MA: MIT Press.

Ruthmann, A., and R. Mantie (2017). *The Oxford Handbook of Technology and Music Education*. New York: Oxford University Press.

Sachs, C. (1940). *The History of Musical Instruments*. New York: W. W. Norton & Company.

Sarath, E. W., D. E. Myers, and P. S. Campbell (2016). *Redefining Music Studies in an Age of Change: Creativity, Diversity, and Integration*. New York: Taylor & Francis.

Savage, R. W. H. (2017). *Music, Time, and Its Other: Aesthetic Reflections on Finitude, Temporality, and Alterity*. Oxon: Routledge.

Schaeffer, P. (1967). *Solfège de l'objet Sonore*. Paris: INA/GRM.

Schomaker, L. (1995). A taxonomy of multimodal interaction in the human information processing system. Technical Report. A report of the ESPRIT Project 8579 MIAMI. Nijmegen, Netherlands: NICI Institute.

Schoonderwaldt, E. (2009). *Mechanics and Acoustics of Violin Bowing: Freedom, Constraints and Control in Performance*. PhD diss., KTH Royal Institute of Technology, Stockholm.

Schütz-Bosbach, S., and W. Prinz (2007). Perceptual resonance: Action-induced modulation of perception. *Trends in Cognitive Sciences 11*(8), 349–355.

Scott, D. B. (Ed.) (2009). *The Ashgate Research Companion to Popular Musicology*. Oxon: Routledge.

Serafin, S. (2004). *The Sound of Friction: Real-Time Models, Playability and Musical Applications*. PhD diss., Stanford University.

Sethares, W. A. (2007). *Rhythm and Transforms*. London: Springer.

Seznec, Y. (2019). *The Book of Knowledge of Impractical Musical Devices*. https://www.impracticaldevices.com/. Accessed March 1, 2022.

Shaffer, L. H. (1989). Cognition and affect in musical performance. *Contemporary Music Review* *4*(1), 381–389.

Shapiro, L. A. (2019). *Embodied Cognition* (Second ed.). Oxon: Routledge.

Shepard, B. K. (2013). *Refining Sound: A Practical Guide to Synthesis and Synthesizers*. New York: Oxford University Press.

Skjulstad, J. (2016). *Circuit Bending as an Aesthetic Phenomenon*. Master's thesis, University of Oslo.

Skogstad, S., K. Nymoen, and M. E. Høvin (2011). Comparing Inertial and Optical MoCap Technologies for Synthesis Control. In *Proceedings of Sound and Music Computing*, Padova, Italy, pp. 421–426.

Skogstad, S., K. Nymoen, Y. D. Quay, and A. R. Jensenius (2011). OSC Implementation and Evaluation of the Xsens MVN Suit. In *Proceedings of the International Conference on New Interfaces for Musical Expression*, Oslo, pp. 300–303.

Small, C. (1998). *Musicking: The Meanings of Performing and Listening*. Music/Culture. Hanover, NH: University Press of New England.

Smalley, D. (1997). Spectromorphology: Explaining sound-shapes. *Organised Sound* *2*(02), 107–126.

Smallwood, S. (2017). Author Commentary: The Hemispherical Speaker and Beyond. In A. R. Jensenius and M. J. Lyons (Eds.), *A NIME Reader: Fifteen Years of New Interfaces*. Berlin: Springer.

Smith, J. (1992). Physical modeling using digital waveguides. *Computer Music Journal* *16*(4), 74–91.

Smith, S. (2013). *Hip-Hop Turntablism, Creativity and Collaboration*. Burlington, VT: Ashgate.

Smus, B. (2013). *Web Audio API*. Beijing: O'Reilly Media.

Snell, J. M. (1983). Sensors for Playing Computer Music with Expression. In *Proceedings of the International Computer Music Conference*. Rochester, NY, pp. 113–126.

Snyder, B. (2000). *Music and Memory: An Introduction*. Cambridge, MA: MIT Press.

Solberg, R. T., and A. R. Jensenius (2017). Pleasurable and Intersubjectively Embodied Experiences of Electronic Dance Music. *Empirical Musicology Review* *11*(3-4), 301–318.

Solberg, R. T., and A. R. Jensenius (2019). Group behaviour and interpersonal synchronization to electronic dance music. *Musicae Scientiae 23*(1), 111–134.

Sporrel, K., S. R. Caljouw, and R. Withagen (2017). Gap-crossing behavior in a standardized and a nonstandardized jumping stone configuration. *PLOS ONE 12*(5), e0176165.

Stember, M. (1991). Advancing the social sciences through the interdisciplinary enterprise. *The Social Science Journal 28*(1), 1–14.

Stenslie, S. (2011). *Virtual Touch: A Study of the Use and Experience of Touch in Artistic, Multimodal and Computer-Based Environments*. PhD diss., Oslo School of Architecture and Design.

Stern, D. N. (2004). *The Present Moment in Psychotherapy and Everyday Life*. New York: WW Norton & Company.

Sterne, J. (2003). *The Audible Past: Cultural Origins of Sound Reproduction*. Durham, NC: Duke University Press.

Streeck, J., C. Goodwin, and C. LeBaron (2011). *Embodied Interaction: Language and Body in the Material World*. Cambridge, UK: Cambridge University Press.

Sudnow, D. (1993). *Ways of the Hand: The Organization of Improvised Conduct*. Cambridge, MA: MIT Press.

Sudnow, D. (2001). *Ways of the Hand*. Cambridge, MA: MIT Press.

Sullivan, J., and M. M. Wanderley (2019). Surveying Digital Musical Instrument Use Across Diverse Communities of Practice. In *International Symposium on Computer Music Multidisciplinary Research*, Marseille, France.

Swarbrick, D., D. Bosnyak, S. R. Livingstone, J. Bansal, S. Marsh-Rollo, M. H. Woolhouse, and L. J. Trainor (2019). How Live Music Moves Us: Head Movement Differences in Audiences to Live Versus Recorded Music. *Frontiers in Psychology 9*(2682).

Sweller, J. (1994). Cognitive load theory, learning difficulty and instructional design. *Learning and Instruction 4*, 295–312.

Tahiroglu, K., M. Gurevich, and R. B. Knapp (2018). Contextualising Idiomatic Gestures in Musical Interactions with NIMEs. In *Proceedings of the International Conference on New Interfaces for Musical Expression*, Blacksburg, VA, pp. 126–131. Virginia Tech.

Tanaka, A. (1993). Musical technical issues in using interactive instrument technology with application to the BioMuse. In *Proceedings of the International Computer Music Conference*, Waseda, Japan, pp. 124–124.

Tatar, K., and P. Pasquier (2019). Musical agents: A typology and state of the art towards Musical Metacreation. *Journal of New Music Research 48*(1), 56–105.

Théberge, P. (1997). *Any Sound You Can Imagine: Making Music/Consuming Technology*. Hanover, NH: Wesleyan University Press.

Thelle, N. J. W. (2010). *Making Sensors Make Sense: Challenges in the Development of Digital Musical Instruments*. Master's thesis, University of Oslo.

Thomas, N. J. (2007). Mental Imagery. In E. N. Zalta (Ed.), *The Stanford Encyclopedia of Philosophy*. The Metaphysics Research Lab, Stanford University. http://plato.stanford.edu/archives/win2003/entries/mental-imagery. Accessed April 8, 2009.

Thompson, W. F., P. Graham, and F. A. Russo (2005). Seeing music performance: Visual influences on perception and experience. *Semiotica 2005*(156).

Thoresen, L., and A. Hedman (2015). *Emergent Musical Forms: Aural Explorations*. Department of Music Research and Composition, University of Western Ontario.

Timmers, R., F. Bailes, and H. Daffern (Eds.) (2022). *Together in Music: Participation, Co-Ordination, and Creativity in Ensembles*. Oxford: Oxford University Press.

Traer, J., and J. H. McDermott (2016). The perception of reverberation is constrained by environmental statistics. *The Journal of the Acoustical Society of America 139*(4), 2210–2210.

Trevarthen, C. (1999). Musicality and the intrinsic motive pulse: Evidence from human psychobiology and infant communication. *Musicae Scientiae 3*(1), 155–211.

Trueman, D. (2007). Why a laptop orchestra? *Organised Sound 12*(2), 171–179.

Tsay, C.-J. (2013). Sight over sound in the judgment of music performance. *Proceedings of the National Academy of Sciences 110*(36), 14580–14585.

Tsoukalas, K., J. Kubalak, and I. I. Bukvic (2018). L2OrkMote: Reimagining a Low-Cost Wearable Controller for a Live Gesture-Centric Music Performance. In *Proceedings of the International Conference on New Interfaces for Musical Expression*, Blacksburg, VA, pp. 275–280. Virginia Tech.

Turchet, L., T. West, and M. M. Wanderley (2019). Smart Mandolin and Musical Haptic Gilet: Effects of vibro-tactile stimuli during live music performance. In *Proceedings of Audio Mostly*, Nottingham, UK.

van Eck, C. (2017). *Between Air and Electricity: Microphones and Loudspeakers as Musical Instruments*. New York: Bloomsbury Academic.

Van Nort, D. (2010). *Modular and Adaptive Control of Sound Processing*. PhD diss., McGill University.

Vines, B., C. Krumhansl, M. Wanderley, and D. Levitin (2006). Cross-modal interactions in the perception of musical performance. *Cognition 101*(1), 80–113.

Volpe, G., A. D'Ausilio, L. Badino, A. Camurri, and L. Fadiga (2016). Measuring social interaction in music ensembles. *Philosophical Transactions of the Royal Society B: Biological Sciences 371*(1693), 20150377.

von Hornbostel, E. M., and C. Sachs (1914). Systematik der musikinstrumente. Ein versuch. *Zeitschrift für Ethnologie 46*(H. 4/5), 553–590.

von Hornbostel, E. M., and C. Sachs (1961). Classification of Musical Instruments: Translated from the Original German by Anthony Baines and Klaus P. Wachsmann. *The Galpin Society Journal 14*, 3–29.

Voulodimos, A., N. Doulamis, A. Doulamis, and E. Protopapadakis (2018). Deep Learning for Computer Vision: A Brief Review. *Computational Intelligence and Neuroscience 2018*, e7068349.

Waisvisz, M. (1985). The Hands, a set of remote MIDI-controllers. In *Proceedings of the International Computer Music Conference*, San Francisco, pp. 313–318. International Computer Music Association.

Waisvisz, M. (2004). Crackle history. http://www.crackle.org/CrackleBox.htm. Accessed March 1, 2022.

Wanderley, M. M. (2001). *Performer-Instrument Interaction. Application to Gestural Control of Sound Synthesis*. PhD diss., University Paris VI.

Wanderley, M. M., and N. Orio (2002). Evaluation of input devices for musical expression: Borrowing tools from hci. *Computer Music Journal 26*(3), 62–76.

Wanderley, M. M., and P. Depalle (2004). Gestural control of sound synthesis. *Proceedings of the IEEE 92*(4), 632–644.

Wanderley, M. M., P. Depalle, and O. Warusfel (1999). Improving instrumental sound synthesis by modeling the effects of performer gesture. In *Proceedings of the International Computer Music Conference*, Beijing, pp. 418–421.

Wang, G. (2018). *Artful Design: Technology in Search of the Sublime*. Redwood City, CA: Stanford University Press.

Wang, S., M. M. Wanderley, and G. Scavone (2020). The Study of Mapping Strategies Between the Excitators of the Single-Reed Woodwind and the Bowed String. In H. Li, S. Li, L. Ma, C. Fang, and Y. Zhu (Eds.), *Proceedings of the 7th Conference on Sound and Music Technology (CSMT)*, Volume 635, pp. 107–119. Singapore: Springer.

Warren, W. H., and R. R. Verbrugge (1984). Auditory Perception of Breaking and Bouncing Events: A Case Study in Ecological Acoustics. *Journal of Experimental Psychology 10*(5), 704–712.

Weiner, I. B. (2003). *Handbook of Psychology, Experimental Psychology*. New York: John Wiley & Sons.

Weium, F., and T. Boon (Eds.) (2013). *Material Culture and Electronic Sound*. Volume 8. Washington, DC: Smithsonian Institution Scholarly Press.

Wenger, E. (1998). *Communities of Practice: Learning, Meaning, and Identity*. Cambridge, MA: Cambridge University Press.

Wessel, D., and M. Wright (2001). Problems and Prospects for Intimate Musical Control of Computers. In *Proceedings of the International Conference on New Interfaces for Musical Expression*, Seattle, pp. 11–14.

Wessel, D., and M. Wright (2002). Problems and prospects for intimate musical control of computers. *Computer Music Journal 26*(3), 11–22.

Williams, A., V. Witani, J. Westendorp, and A. White (2011). Network devices, methods and/or systems for use in a media network. Patent: EP2255541A4. U.S. Patent Application No. 12/735,925. 2011 Jan 6.

Wilson, M., and G. Knoblich (2005). The Case for Motor Involvement in Perceiving Conspecifics. *Psychological Bulletin 131*(3), 460–473.

Winters, R. M., A. Savard, V. Verfaille, and M. M. Wanderley (2012). A Sonification Tool for the Analysis of Large Databases of Expressive Gesture. *The International Journal of Multimedia & Its Applications* 4(6), 13–26.

Withagen, R., and S. R. Caljouw (2017). Aldo van Eyck's Playgrounds: Aesthetics, Affordances, and Creativity. *Frontiers in Psychology* 8, 1130.

Withagen, R., H. J. de Poel, D. Araújo, and G.-J. Pepping (2012). Affordances can invite behavior: Reconsidering the relationship between affordances and agency. *New Ideas in Psychology* 30(2), 250–258.

Wöllner, C., D. Hammerschmidt, and H. Albrecht (2018). Slow motion in films and video clips: Music influences perceived duration and emotion, autonomic physiological activation and pupillary responses. *PLOS ONE* 13(6), e0199161.

Wright, M. (2002). Problems and prospects for intimate and satisfying sensor-based control of computer sound. In *Sensors and Input for Multimedia Systems*, Santa Barbara, CA.

Wright, M. (2017). Unsolved Problems and Continuing Prospects for Intimate Musical Control of Computers. In A. R. Jensenius and M. J. Lyons (Eds.), *A NIME Reader: Fifteen Years of New Interfaces*, pp. 22–24. Berlin: Springer.

Wright, M., and A. Freed (2017). Open Sound Control: Some Context and Reflections on Thirteen Years' Advances. In A. R. Jensenius and M. J. Lyons (Eds.), *A NIME Reader: Fifteen Years of New Interfaces*, pp. 140–142. Berlin: Springer.

Wright, M., and A. Freed (1997). Open SoundControl: A New Protocol for Communicating with Sound Synthesizers. In *International Computer Music Conference*, Thessaloniki, Greece, pp. 101–104.

Wright, M., A. Freed, A. Lee, T. Madden, and A. Momeni (2001). Managing complexity with explicit mapping of gestures to sound control with osc. In *International Computer Music Conference*, Havana, Cuba, pp. 314–317.

Wurtzler, S. (2007). *Electric Sounds: Technological Change and the Rise of Corporate Mass Media*. Film and Culture Series. New York: Columbia University Press.

Xambó, A. (2018). Who Are the Women Authors in NIME?–Improving Gender Balance in NIME Research. In *Proceedings of the International Conference on New Interfaces for Musical Expression*, Blacksburg, VA, pp. 174–177. Virginia Tech.

Yost, W. A., A. N. Popper, and R. R. Fay (Eds.) (2008). *Auditory Perception of Sound Sources*. New York: Springer.

Zeigler, E. F. (1990). Don't forget the profession when choosing a name. In *The Evolving Undergraduate Major*. Champaign, IL: Human Kinetics.

Zelechowska, A., V. E. González Sánchez, B. Laeng, and A. R. Jensenius (2020). Headphones or Speakers? An Exploratory Study of Their Effects on Spontaneous Body Movement to Rhythmic Music. *Frontiers in Psychology* 11, 698.

Index